ASTRO REALITY

AVOIDING CONFLICT IN OUTER SPACE.

ROY H SACH

ECHO BOOKS

First published in 2015 by Barrallier Books Pty Ltd,
trading as Echo Books

Registered Office: 35-37 Gordon Avenue, West Geelong, Victoria 3220, Australia.

www.echobooks.com.au

National Library of Australia Cataloguing-in-Publication entry (pbk)

Creator: Sach, Roy, author.

Title: Astro reality : avoiding conflict in outer space/Roy Sach.

ISBN: 9780994281135 (paperback).

Subjects: Space warfare--Prevention. Space control (Military science). Space pollution. Space weapons (International law). Outer space--International cooperation.

Dewey Number: 358.8

Book and cover design by Peter Gamble, Ink Pot Graphic Design, Canberra.

Set in Garamond Premier Pro 12/17 and Minerva small caps.

www.echobooks.com.au

Dedicated with respect and admiration to
Professor Malcolm Ross Walter
1944 -

CONTENTS

Tables and Figures

Acronyms and Symbols

Acronym/Symbol	Meaning
aka	also known as
ANGELS	Autonomous Nanosatellite Guardian for Evaluating Local Space
ANZUS	Australia, New Zealand and United States (security treaty between)
ASAT	anti-satellite
BCE	before common era
BMD	Ballistic Missile Defence
c	speed of light
c.	circa, approximately
°C	Celsius temperature in degrees
CE	common era
COPUOS	Committee on the Peaceful Uses of Outer Space
CW	continuous wave
DPRK	Democratic People's Republic of Korea
EAGLE	Evolutionary Air and Space Global Laser Engagement
EGM	Earth Gravitational Model
ESA	European Space Agency
EU	European Union
FEL	free-electron laser
g	Magnitude of acceleration of a body due to gravitational attraction of Earth. ('g' is also the formal symbol for gram. The distinction will be contextual.)
GA	General Assembly

Acronym/Symbol	Meaning
GDP	Gross Domestic Product
GEO	geosynchronous orbit
GLONASS	*Globalnaya Navigatsionnay Sistema*
GNSS	Global Navigation Satellite System
GPS	Global Positioning System
Gy	gray (used in the context of absorbed radiation dose)
IAEA	International Atomic Energy Agency
ICAO	International Civil Aviation Organisation
ICBM	intercontinental ballistic missile
ISS	International Space Station
ITU	International Telecommunications Union
KEW	Kinetic Energy Weapon
Kt	kiloton
LEO	Low Earth Orbit
Mach number	Ratio of relative speeds of a fluid and a rigid body to the speed of sound in that fluid with identical temperature and pressure.
MEO	mid-Earth, or medium-Earth orbit
MIRACL	Mid-Infrared Advanced Chemical Laser
MIRV	multiple independently-targeted re-entry vehicle
MOPA	master oscillator power amplifier
Mt	megaton
MTCR	Missile Technology Control Regime
NASA	National Aeronautics and Space Administration
NASDAQ	National Association of Securities Dealers Automated Quotations
NATO	North Atlantic Treaty Organisation
NDRE	nuclear-detonation-induced radiation effects
Nd:YAG	neodymium-doped yttrium-aluminium-garnet
NPT	Non-Proliferation Treaty
PAROS	Prevention of Arms Race in Outer Space
PEM	polymer electrolyte membrane cell
PRC	People's Republic of China
QZSS	Quasi-Zenith Satellite System
RMS	root mean square value (in the physics context)

Acronym/Symbol	Meaning
RoK	Republic of Korea
S&P 500	Standard & Poor's 500
SI	*Systeme Internationale d'Unites*
SLAC	Stanford Linear Accelerator Center
SLBM	submarine-launched ballistic missile
SM-3	Standard Missile Type 3
THAAD	Terminal High Altitude Area Defense
UAV	unpopulated aerial vehicle
UN	United Nations
US	United States of America
USAF	United States Air Force
USD	dollar currency of United States of America
USSR	Union of Soviet Socialist Republics
V-2	*Vergeltungswaffe* (German missile)
XSS	Experimental Spacecraft System

Conventions

Nation States

Names or abbreviations used in this work for nation States reflect political chronology. For example, 'Russia' was applied to that State when referring to events before and including 1917; between 1918 and 1991 'Union of Soviet Socialist Republics (USSR)' was adopted, noting the Union incorporated more States than did the former Russia; and for events after 1991 'Russia' was again used. Similarly, the name 'China' was used in the context of events prior to and including 1949, and 'People's Republic of China (PRC)' for events after that year.

The terms 'nation State', or 'State', were used in preference to 'country', 'homeland' and similar terms. 'State' and 'nation State' reflect terminology commonly employed in formal documents such as treaties.

If the name of a nation State extended to two or more words, abbreviations were adopted; examples being: United States of America (US), Democratic People's Republic of Korea (DPRK) or Republic of Korea (RoK).

Names of Organisations

For organisations, a convention similar to that applied to nation States was used. The United Nations became 'UN' and European Space Agency was abbreviated to 'ESA'.

Measurements, all Categories

With three exceptions, terminology related to measurements was used, defined and abbreviated in accordance with conventions adopted by the *Systeme Internationale d'Unites* (SI), including the SI derived and combined units listed by the *Bureau International des Poids et Mesures* <http://www.bipm.org/en/si/si_brochure/chapter4/table7.html>. These abbreviations were included without elaboration or explanation.

The first of the three exceptions was made to reflect terminology used almost invariably in aviation and space- related literature. In that context the symbol '*g*' was employed in reference to the magnitude of acceleration of a body. The *Systeme Internationale d'Unites* had determined the formal abbreviation for that unit to be ' m/s^2 '.

The second and third exceptions were that the terms 'megaton' and 'kiloton' were employed in relation to nuclear weapons. The *Systeme Internationale d'Unites* does not recognise megaton, kiloton, megatonne or kilotonne in the context of nuclear weapons.

Throughout the manuscript, mass expressed in 'tonnes' should be interpreted as generally indicative. This is because data was commonly obtained from sources referring to 'tons' , often without discrimination between so-called short tons (2 000 pounds) and long tons (2 240 pounds). In cases where it was not possible to ascertain the category of ton intended in source material, the long or 2 240 pound ton was assumed in conversions to tonnes.

Nuclear Weapons

'Nuclear weapons' and similar published phrases seldom discriminate between weapons commonly referred to as 'atomic' bombs, in which a fission process occurs, and those known as 'hydrogen' or 'thermonuclear' bombs which employ a dominantly fusion-based process. This work addressed actual and potential damage to spacecraft from detonation of nuclear weapons and was not concerned over whether the process involved fission or fusion.

Gender Neutral Language

General neutral language was used even if it contrasted with common usage. For example, '*unmanned* aerial vehicles' were identified as '*unpopulated* aerial vehicles' (UAVs).

References to Internet Websites

Most references to internet websites provided details only to the primary identifier level (eg. <http://www.xyz.com>). It was, and is likely to remain, faster and more efficient to locate specific items cited by utilising available key words and search engines than to pursue full identification details in each case. Also, website redesigns can result in the repositioning of information.

Financial Currency

Financial currency was shown in United States of America(n) dollars, expressed as 'USD'. In most examples, the chronological price base for currency was not included in source material consulted, and therefore the financial estimates or records shown were merely indicative. No requirements to convert currencies arose and so conversion rates were not addressed.

Foreword

A common question at conferences around the world is, what would life be like without space. It's not a parlour game question but a serious attempt to comprehend what we would do if we didn't have access to space. Much of our modern way of life is enabled by space and it is hard to imagine life without global navigation systems, weather forecasts informed by space observations, support to banking and business enabled by satellite communications and for some satellite television.

However, access to space is not guaranteed. The greatest threat to our continued use of space comes from the militarisation of space. This process seems inexorable. Even now many military capabilities are either enabled by space based systems or need to travel through space to reach their target. Harder to substantiate but entirely feasible is the presence of weapons systems already in space designed to attack terrestrial targets or destroy or disable enemy space based capabilities.

Even though the prospect of life without space is hard to imagine the history of mankind suggests that we will find a way to spoil the party. The prospect of space becoming a zone of conflict is very real. In the 20th Century space was named as a new domain of warfare after land, sea and air. Space is the ultimate 'high ground' of warfare. Control of space by any nation, if that were possible, would bring significant power.

There would be no winners if there were to be a war in space. The consequences of losing access to space are so great that we need to find ways to avoid the possibility. Even without battles waged in space the current largely uncontrolled use of space is a significant threat as the proliferation of space junk from out of control or discarded satellites makes it harder to use space safely. We have polluted the earth and the oceans and now we are polluting space.

Finding ways of preventing conflict in space won't be easy and will require imagination, tolerance and patience. It will require a multi-disciplinary approach involving scientific, diplomatic, legal, and economic means. As a global community we urgently need to explore all means to limit the threat of war in space. The prospects aren't good. As space becomes more and more crowded and more countries seek independent access to space global cooperation is essential. At the moment there is no clear pathway to suggest that productive international agreements on the use of space will be achieved soon.

Roy Sach has presented a compelling case that something has to be done. His work is extensive and exhaustive and he has ranged widely across geopolitics, history, science and technology. His judgments are sound and he deserves to be listened to. His work provides a timely warning that action is required if mankind is to continue gaining the benefits of space as a means of improving our lives and wellbeing.

Professor Peter Leahy AC
Lieutenant General (Retd)
University of Canberra

Acknowledgements

I am indebted to many people who shared with me their time and expertise while I was investigating issues considered in this manuscript.

In particular, I am grateful for academic and professional advice provided by: Professor M Walter, Professor (Lieutenant General) P Leahy, Professor C Rizos, Professor J Kennewell, Professor S Freeland, Professor (Commodore) S Bateman, Professor R Boyce, Associate Professor J Bailey, Doctor D Leary, Doctor J Horner, and Doctor T Williams. Other people who offered valuable support include: Mister B Biddington, Ms J Lewis, Doctor N Siemon, Ms K Dougherty, Mister G Roberts, Doctor J O'Rourke, Ms K Jones, Ms P Hamilton, Doctor T Ferrara, Doctor N Gambale, Doctor B Cardillo, Professor R Huisken, Doctor C Oliver and Doctor J Fergusson. The IT Customer Services staff at University of New South Wales, Australia, provided exceptionally fine service as did staff at the integrated libraries.

My wife, originally Geertje Martha Huigens, must also be mentioned for I could not have undertaken this work without her consistent support. She was two years-old when Germany invaded the Netherlands in 1940, beginning a period of foreign domination lasting until she turned seven. Food was scarce during The Occupation, and after her parents began concealing Jewish refugees a meagre pantry had to feed yet more people. Later, when suspicious German soldiers began raiding the family home to search for Jews they would interrogate this vulnerable child in isolation from other family members. But the confused and undernourished little girl disclosed nothing of incriminating value and her family, together with their Jewish guests, survived the Second World War. From that crucible emerged a compassionate, refined, linguistically and musically talented lady. She committed decades of her life to supporting both my foibles and my ambitions, often at the expense of opportunities to pursue her own objectives.

RHS

Canberra, 2015

INTRODUCTION

In this book I review potential warfare in outer space from strategic, legal and technologically analytical perspectives. My choice of title, *Astro-reality: avoiding conflict in outer space,* owes much to Everett C Dolman's 2002 work, *Astropolitik: Classical geopolitics in the space age.*[1]

When reading *Astropolitik* I kept wondering how Everett Dolman could have reached conclusions so fundamentally different to mine, despite our obvious awareness of many identical or closely related issues. Nevertheless, if we met I am confident we would agree on several core issues.

We would have no doubt the capacity to influence space-sourced data and its associated services can enhance international prestige while conferring geopolitical power. Humanity's utilisation of outer space therefore has implications for international relations and global stability.

We would agree most activities in outer space have been enabled by a small number of mutually competitive States. The much larger cohort of other States, some with small numbers of spacecraft, has become critically dependent on space-sourced data and services often provided by technologies they neither own nor control and over which they exert minimal influence.

Dolman's study of space-related geopolitics and my analysis of the attributes of warfare both point to the same challenge; there is a realistic prospect of warfare in outer space and it poses a grave threat to all nation States. But from that point onwards our deliberations and conclusions diverge to follow irreconcilable paths.

In *Astropolitik,* Everett Dolman's engaging style, his international reputation and the compatibility of his conclusions with attitudes common in the United States of America

(US) combine to produce a potent and overtly appealing argument. There is no doubt in my mind where the mass and volume of judgmental opinion would fall if our books were compared.

However, I intend to persist because knowledge is not advanced by concurring with established viewpoints. Readers who also intend persisting are cautioned not to expect from me anything approaching the depth of Dolman's impressive research into theories of international relations and similar issues. Forty-two years of defence service, both in uniform and as a civilian, scoured the subtleties from my pen. My work is therefore constrained by advice attributed to Albert Einstein; *Everything should be made as simple as possible, but not simpler.*

In this context I considered the likelihood of conflict occurring in outer space and identified ways to reduce associated threats to global security out to the year 2030 CE, approximately. I also examined tactics and strategies with potential to contain or limit the terrestrial harm likely to be caused by any such conflict.

Investigating the *likelihood* of warfare in outer space might seem unnecessary given my previous claim, with which Everett Dolman would agree, there is a realistic prospect of such an event happening. I nevertheless revisited it because some writers still cling to notions that outer space is or can become a zone of peace and tranquility.

I have accepted ~2030 CE as roughly the limit of foreseeability for defensible, evidence-based analyses of issues explored. From accessible information I concluded the development of space-related weapons typically occurs over several decades during which progress toward operational deployment can often be observed. However, the potential for unanticipated technologies to emerge and exert strategic influence must increase over time. Limiting this investigation to ~ 2030 CE helped constrain that risk without eliminating it.

While conclusions about technological progress to 2030 incorporate some uncertainty, assessments of future geopolitical developments are likely to be even less reliable. Globally significant events such as the emergence of a nuclear-armed State of Israel, the collapse of the Union of Soviet Socialist Republics (USSR), the 2001 attacks on the United States of America (US) and the Global Financial Crisis of 2007 - 2008 were not widely anticipated and certainly not anchored chronologically. In this work the inevitable uncertainties associated with prospective geopolitical developments have been diluted by identifying broad and consistent trends in the conduct of warfare and by demonstrating how those activities evolved to affect outer space.

Space warfare could potentially influence many and possibly most aspects of terrestrial activity, and *vice versa*. Boundaries to this investigation were therefore necessary. I applied the following constraints:

a. Warfare in Outer Space Defined. Warfare in outer space was considered to be, *Hostile action which could or does threaten, disrupt or destroy the operation of any human-made object functioning in outer space and either: (a) orbiting Earth or another celestial object, (b) in transit between orbits or between celestial objects, or, (c) located on a celestial object other than Earth.*

b. Containment Defined. Containment strategies and tactics were recognised as, *Laws, procedures, policies, actions or technologies with potential to either reduce the likelihood of conflict occurring in the outer space region, or to reduce the consequences of such conflict if it occurs.*

c. Activities Excluded. Space-to-Earth warfare was not addressed because it had already been examined in research related to missile defence and through scholarly responses to US President Reagan's strategic defence initiative, which he announced in March 1983. Earth-to-Earth warfare is by definition a component of terrestrial warfare, notwithstanding such activity could damage or destroy space-related facilities such as launch platforms or the ground stations used to control spacecraft. Cyber warfare was not explored because distinctions between outer space and the terrestrial environment are not fundamental to the manner in which cyber warfare is conducted.

d. Esoteric Possibilities Ignored. I ignored technologies and tactics I considered esoteric and impracticable as components of warfare in outer space to the year 2030. These included the potential to use asteroids as weapons as well as theories addressing destructive uses of antimatter. Similarly, the prospect of conflict in or over the Lagrange 'points' as components of space warfare appeared inconceivable during the period under review.

I have not considered in detail the political, economic and other consequences of space-based data and services being denied or limited following warfare in outer space. Those topics have been explored comprehensively through the published proceedings of conferences and in journals employing the '*day without space*' theme. Reiteration would have added nothing of value.

Although this book is focused on outer space there is no widely acknowledged altitude at which the dense terrestrial atmosphere can be regarded as terminating and outer space as beginning. Appendix 1, *Identification of the Outer Space Boundary*, argues for the boundary to be accepted as 80 km, linear and vertical, above the datum Earth Gravitational Model 2008. I applied that boundary consistently.

I presented strategies and tactics to contain or discourage warfare in outer space under three categories. *Preserve Freedom* considers the policies and initiatives of major spacefaring States, providing a framework that cannot safely be ignored by other entities. *Acquire Freedom* reviews opportunities for other States to enhance their own security by reducing the potential for warfare to occur in outer space. *Share Freedom* explores opportunities for global cooperation aimed at making the outer space region more secure for current and future generations.

The book is based on a manuscript I submitted to the University of New South Wales (UNSW), Australia. It is accessible through the University's main library where it is held electronically (R H Sach, *Containing space warfare in the early decades of the twenty-first century*). To compress this complicated topic into a 100 000-word PhD thesis I defaulted to using endnotes, thousands of them. It was neither practicable nor desirable to republish them in a book of this nature. Therefore, readers wishing to further explore specific issues, or my justifications for them, will need to consult the manuscript in the UNSW Main Library.

I abhor unauthorised disclosures involving classified official information and will neither cite them nor quote from them. When compiling both the thesis and this book I also checked periodically to eliminate any possible inclusion of data or information to which, in the past, I might have had confidential, commercial or other privileged access from any source.

The book is not an exercise in fanciful idealism. Proposals to contain space warfare have been anchored to relevant information supported in most cases by established precedent. Given appropriate international cooperation and political-will, the proposals could be implemented. Most potential strategies identified are not interdependent. Progress made with any of them could contribute to enhanced security of the outer space region.

Chapter 1. The Foundations of Warfare in Outer Space

The first human-made object to enter outer space was probably a German *Vergeltungswaffe-2* (V-2) missile launched during the Second World War. It is not known whether the missile's chief architect, Wernher von Braun, either knew or cared if his invention could enter a region regarded as outer space. But there is no doubt following Germany's surrender, in May 1945, US and USSR personnel raced to obtain records of German missile technology, later incorporating it into their own nascent military-space programs.

During the Cold War between the US and USSR (1945–1991) both primary belligerents adapted and evolved German technologies to build nuclear-armed long-range missiles with which to attack each other. By May 1957 the USSR was able to launch a missile with a range of 8,000 km and an apogee (maximum altitude) of 1,000 km. One of the earliest American space policy documents, *US policy on outer space*, 20 June 1958, listed acquisition of both ballistic missiles and anti-ballistic missiles as weapons either planned or in immediate prospect. By then, some such weapons had already entered service and more were to follow.

The US and USSR were soon forced to acknowledge that warfare involving these nuclear weapons could result in mutual annihilation and safeguards were needed to guard against false alarms or other errors. Ironically, the advent of long-range, nuclear-armed missiles therefore contributed in some respects to *reducing* the likelihood of either State initiating a strategic nuclear attack. The precedent for an enduring and ambivalent relationship between humanity and outer space was thus also established.

This relationship became yet more complicated after the USSR used a modified missile to launch the very first artificial satellite, *Sputnik 1* on 4 October 1957. The *Sputnik 1* launch demonstrated some missiles could be reconfigured to function in a potentially less aggressive role as launch vehicles. The 1958 *US policy on outer space* identified a range of beneficial scientific and related applications for artificial satellites. A contemporary USSR publication, *Soviet Space Science* listed similar applications for Soviet spacecraft.[1]

Emerging opportunities to provide and control the space-sourced data and services desired by other States offered new avenues by which the two prominent global entities could increase their international prestige, power and influence. Space-sourced data and services were expanded to support trade, agriculture, finance, communications and other categories of terrestrial activity. Science became increasingly space-reliant as instruments in outer space provided data for applications as diverse as meteorology, environmental monitoring and astronomy. A near addiction to space data spread to commercial and other entities.

Military requirements for space-sourced data and services also increased, reaching a level where armed forces were compelled to utilise non-military spacecraft to compensate for shortfalls in their dedicated communications, meteorology and navigation capabilities. The high financial cost of providing space-based services further motivated States to launch dual-use military and non-military spacecraft. Nevertheless, specialised defense or intelligence-related spacecraft were retained, and this seems likely to persist indefinitely.

While that medley of trends was emerging, during the 1950s and 60s attempts were being made to establish the foundations for global cooperation in outer space. The results of those endeavours are now embodied substantially in the *Treaty on Principles Governing the Activities of States in the Exploration and Use of Outer Space, including the Moon and Other Celestial Bodies* (the Outer Space Treaty, 1967).

Given the evolution I have described, the contemporary benefits of safeguarding space-sourced data and services seem so apparent that warfare in outer space should be inconceivable. However, perceptions of potential to expand their military and economic authority have often proved an irresistible temptation for powerful States. And States are accustomed to protecting valuable national assets. Accordingly, some have been developing and testing weapons with which to fight in outer space.

Typically, past claims that warfare in outer space is a realistic threat have focussed on an assumption that the above description of events supplied an adequate foundation from which further investigations might proceed. The assumption is dangerous.

The so-called *space age*, the second half of the 20th century, saw increasingly assertive independence of States in the Middle-East and on the South American continent, disintegration of the USSR, the rise to greater power and influence of the People's Republic of China (PRC) as well as of India, Japan, Brazil, Israel, Iran, Republic of Korea (RoK) and the Democratic People's Republic of Korea (DPRK). Non-State activism and new organisations dominated by allegedly religious fanatics gained more prominence than formerly. Global economic interdependence increased. Arguably, the consequences of each of those developments, and many others, are still emerging. Lessons from studying the period in which they featured prominently are therefore likely to be tenuous, and the space age is an inadequate timeframe for detecting and analysing geopolitical influences affecting outer space.

Human interest in space as a potential venue for warfare merely represents one further and predictable step in an evolution that has been building for thousands of years. The remainder of this chapter therefore reaches across 12 millennia to consider factors affecting human-to-human violence, each with implications for outer space. Possible limitations to and unpredictable influences operating on such violence are also identified.

This aspect of the investigation deserves a work of encylopedia-like dimensions, but the practical constraints of publication limited examples used to a basic framework selected from a cornucopia of available evidence.

Superior range and lethality of weapons have often emerged as critical prerequisites to military victories. Commanders have pursued these objectives since the dawn of recorded human history. One of the earliest verifiable land battles occurred during c.10 000 BCE to the north of Wadi Halfa in Egypt. Modern excavations have confirmed the use of bows and arrows which applied physical leverage to achieve lethality, increase range and amplify the capacities of human physiques. In China, the more potent crossbow was invented on an unknown date BCE. By c.270 BCE Europeans were building siege engines based on matrices describing the design required to deliver a nominated mass across a specified distance, in a manner broadly similar to the tables used later by armies for artillery control.

Subsequently, gunpowder further increased the range and lethality of some weapons. Records exist of the Chinese using gunpowder for propelling projectiles in 1132 CE and of explosive grenades employed during a battle in 1161 CE. Knowledge of the relevant chemistry either migrated to or was discovered independently in other locations.

Cannon changed ships from seagoing transports, and occasionally battering rams, into weapons platforms. By 1697 the English navy alone was equipped with 9 912 cannon. In 1916 artillery could be fired with such precision it was possible to design a '*creeping barrage*' in which projectiles were aimed to clear a path for waves of advancing troops.

By the 17th century, humanity's quest for increased range had extended into the first totally alien environment. Cornelius Jacobszoon Drebbel (1572 – 1633) built a vessel designed to travel beneath the surface of water, a place in which unprotected humanity could not survive. The evolution of these devices continued and by the First World War submarines commonly had operating ranges, surfaced, of some 435 km. They could fire torpedoes at distances of ~ 750 m from their targets. More recently, the British Astute Class nuclear submarines, first commissioned in August 2010, had an operating capacity of 40 global circumnavigations, without refuelling, and were carrying Tomahawk Block IV missiles with a range of ~1 100 km.

Combat expanded into the air following Clément Ader's demonstration flight in a human-crewed, engine-powered, heavier-than-air machine on 9 October 1890.[2] By 1914 airborne observers were dropping grenades over the sides of fabric-covered flying machines. Specialised fighter and bomber aircraft soon followed. During the Second World War, Boeing's B29 Superfortress bomber had a range of ~5 380 km and a payload of ~9 000 kg, a USSR version being the Tupolev Tu-4 Bull. One of the more modern bomber aircraft is Northrop Grumman's B2 which has a range, without mid-air refuelling, of 12 000 km and a weapons payload, potentially including nuclear weapons, of 18 000 kg.

Modern rockets as weapons have been available since, and possibly before, one version of the Congreve rocket was produced in 1805 with a range exceeding 1 000 m. The weapon was used successfully during 1807 in an attack against Copenhagen, among other military engagements. During the Second World War Germany unveiled the *Vergeltungswaffe-2* (V-2) rocket with a range of 320 km and a 750 kg payload. The V-2, however, was one in a series of planned missiles intended to culminate in a weapon with a 5 150 km range.

These V-series missiles presaged the development of intercontinental ballistic missiles (ICBMs) and, during the 1950s, the US and USSR developed not only the missiles but also the corresponding launch vehicles for spacecraft. Consequently, increasingly potent rockets expanded humanity's access to outer space from low-Earth Orbits (LEOs, ~ 180 – 2 000 km) to mid-Earth orbits (MEOs, ~ 2 000 – ~ 35 000 km) and eventually to geosynchronous orbits (GEOs, ~ 35 780 km) and beyond. (Diagrams illustrating common satellite orbits are at the end of this chapter.)

In turn, ICBMs influenced the design and capabilities of advanced submarines. During the 1960s USS *Halibut*, carried the Regulus missile which had a 45 kiloton (Kt) nuclear warhead with a range of approximately 925 km. During July 2011 *Global Security Newswire* described a PRC-flagged submarine with capacity to launch six nuclear-tipped ballistic missiles each with a range of approximately 8 000 km.[3]

Weapons intended to travel through space before engaging terrestrial targets can now be launched from on the land, on the surface of the sea or beneath the sea surface. These weapons could eliminate almost any terrestrial target with a maximum delay of ~ 90 minutes.[4] Every State is vulnerable. In addition, weapons able to destroy spacecraft can be launched from the land, on the sea surface and possibly from beneath the sea surface, although this latter case does not appear to have been demonstrated.

Since the beginning of recorded history military commanders have also recognised advantages to be gained by the enhanced *mobility* of their troops, their weapons and associated logistics support. During the reign of Ramses III (1186 – 1154 BCE) warriors known as Sea People travelled across water to launch a surprise attack on the Nile Delta region. The Egyptian New Kingdom warriors used horse-drawn chariots to convey them speedily into battles on a major scale for that era. Ancient Romans built a network of hard-surfaced overland roads enabling reliable access to conquered territories. In the period 500 – 1450 CE the mobility and flexibility of manoeuvre available to Mongol cavalry resulted in their Eurasian forces sometimes overwhelming the infantry-centric Byzantine armies.

By 800 BCE floating vessels were being used to expand and secure national strategic and trade interests, leading one author to claim the Aegean had become a Greek sea. But the Greeks did not confine themselves to the Aegean and their trade spread to other areas of the Mediterranean region. Later, European colonial conquests were made possible by sailing vessels capable of voyages lasting weeks or months at a time.

In 1833 a steamship could cross the Atlantic Ocean in 22 days and steamship companies were being established to ply major trade routes. Steamships carried troops in 1861 during the US Civil War. The size and capacities of steamships increased rapidly with tonnages increasing from a common maximum of approximately 5 000 in 1875 to 10 500 by 1890. Maritime mobility by then had evolved to support warfare in every continent except the Antarctic and on every ocean as nations competed for power and influence.

Following the invention of steam-engines, trains contributed to rapid military mobility on land. Railway tracks were established across Europe, Asia, South America and North America during the 19th century. Rail transport aided Prussia's victory in the Seven Weeks' War of 1866.

By the Second World War (1939 – 1945), military mobility incorporated every known form of transport. It included ships, aircraft, trucks, trains and barges together with gliders towed by powered aircraft. A safety device, the parachute, had been converted into a mobility aid allowing troops to be inserted into battlefields from the air.

Long-range missiles later supported and consolidated strategic military mobility. The time lapse between crossing State borders and attacking a geographically remote target could thus be compressed from years, for sword-carrying ancient Romans, to minutes for States with access to the most advanced technologies.

Outer space now offers the most rapid mobility known to military science. Even the hypervelocity vehicles, under development and designed to traverse upper layers of the dense terrestrial atmosphere, cannot match velocities achievable in outer space. Typically a hypervelocity vehicle will function at ~ 2.0 km/s. A spacecraft in a stable orbit at a mean altitude of 35 800 km has a velocity of ~ 3.2 km/s. Spacecraft in lower orbits have yet higher velocities; for example, ~ 7.8 km/s at a mean altitude of 200 km.

I have employed *stealth and deception*, including camouflage, as representative of numerous military tactics with both ancient terrestrial heritage and current utilisation in outer space. Other tactics of similar lineage and contemporary application in space include communications, intelligence collection, surveillance, reconnaissance, coordination of timing, command and control.

Returning to stealth and deception, the Greek myth of the Trojan horse emerged in c.1 200 BCE, illustrating familiarity with these concepts had already developed. In 326 BCE Alexander the Great employed elaborate ruses to confuse his opponents while he was preparing an attack. In the Napoleonic Peninsular War (1808 – 1814) Spanish insurgents used guerrilla tactics to attack the more powerful French army.

During the First World War German combat ships camouflaged as merchant vessels successfully approached and sank allied merchant shipping. In the Second World War, prior to the Second Battle of El Alamein, British forces built dummies of tanks, artillery, trucks and people to draw the attention of opposition forces away from actual preparations for battle. In Europe during June 1944, dummy paratroopers were dropped from the air to the ground. Strips of aluminium foil have been used on many occasions to cause misleading radar signals.

Following the Second World War the US developed radar avoidance for aircraft using a radar–absorbent coating known as Iron Ball. It was applied to the American U2 aircraft until the technique was defeated by more powerful radar interrogation signals. Ships also have used radar reflective or absorbent coatings, examples being Skjold-class patrol boat and the La Fayette-class frigates. Stealth and deception techniques protect radio frequency transmissions.

Stealth and deception are now practised in outer space. Efforts are made through encryption techniques and other radio frequency-related methodologies to prevent unauthorised knowledge of, or interference with, data transmitted by satellites.[5] Some satellites have been designed to minimise the likelihood of detection by radar or visual apparatus. ICBMs can carry dummy warheads and other devices intended to confuse opposition defence technologies.

In parallel with technological progress the strategic *significance of distinctions between the land, sea and air environments*, together with the outer space region, were steadily eroding. This process began subtly with incidents such as the attempted invasion by the Sea People, already mentioned, using the sea surface to convey militia. Similar tactics were used repeatedly over millennia when combatants were either conveyed over water or fought on the surface of enclosed waters, as occurred for example in September 480 BCE when vessels commanded by the Persian Xerxes confronted the Athenian Themistocles' fleet within the Straits of Salamis.

The first submarine to attack a surface ship was the American *Turtle* in 1776. *Turtle* failed to sink any vessel but set a precedent and led to further experiments. By Aug*u*st 1900 the US Navy was able to order six submarines of a viable design. A vessel with similar specifications was then built under licence in Britain. The technology proliferated rapidly as the Dutch, Austrian, German and Russian navies built submarines. Interactions between the sub-surface and surface maritime environments became commonplace as a consequence of submarine warfare during the First World War. Those events led to weapons with anti-submarine capabilities, such as sea-mines, being deployed from other physical environments. Once equipped with long-range missiles, submarines were able to also influence the land environment by firing submarine–launched missiles.

Before the First World War had started aircraft could already bridge the gap between land-based and sea-based operations. In January 1911 Eugene Ely flew the first aircraft to land on a ship.[6] This led to the construction of aircraft carrier ships. Bombers additionally connected the air, land and sea environments by dropping incendiary and explosive devices on land targets and by attacking maritime targets. Airborne transport aircraft consolidated this trend by transporting troops and resupplying forces. Atomic bombing attacks in Japan during 1945 on Hiroshima and Nagasaki demonstrated the potential to use an airborne platform to deliver one weapon with which to destroy the centre of a city.

Following the launch of *Sputnik 1* satellite sensors began collecting State and military intelligence from terrestrial source-targets. Terrestrially-based ICBMs could be used as weapons *in* outer space if they were detonated with the intention of destroying satellites.

After the *Eilat* incident in 1967, when an Israeli destroyer was sunk by missiles launched from Egyptian vessels while still in their harbour, protective missiles fitted to some specialised surface vessels evolved until they could extend to outer space. This capability was demonstrated in February 2008 when a missile launched from the USS *Lake Erie* shot down a US satellite.[7] That sea-based demonstration followed an earlier event in January 2007 when the PRC had used a land-based missile to destroy one of its own satellites.[8]

Interactions between the terrestrial environments of land, sea, air and sub-surface have continued to increase as have the interactions between each of them and outer space. By the end of the 20th century, military land, air and maritime units were interacting with and in varying degrees dependent upon data and services provided by space vehicles. These space-sourced services included: early warning of missile launches and associated missile tracking data, collection of intelligence with strategic and tactical relevance, navigation data, guidance for precise munitions, meteorological services, and provision of broadband data relating to targets and battle damage assessment. Even the long-range unpopulated aerial vehicles (UAVs) are often marionette-like devices linked electronically to satellites transmitting control and navigation data.[9]

Weapons and their related platforms now have the capacity to engage targets on land, in the sea, in the air, beneath the ocean surface and in outer space while operating from any of those environments. From both a tactical and a strategic perspective, all of the terrestrial environments and the outer space region have become interdependent and the traditional boundaries between them are now correspondingly blurred.

Throughout recorded history, technological and logistical influences have compelled military forces to fight using a *combination of contemporary and older weapons*. Before and during the 15th century many cannon could be fired only three times each day; consequently, siege engines often accompanied cannon into battlefields. In 1862, during the US Civil War, the State of Georgia authorised manufacture of 10 000 steel pikes as an alternative to muskets. By the First World War, when battle tanks, trucks, steamships and railway trains had become available, camels, horses and donkeys were used for transport as they had been for millennia. Homing pigeons were also employed to carry messages if electronic communications were unavailable to combatants. A similar trait was evident during the Second World War when Poland deployed 70 000 troops still relying upon horses for their mobility.

The mix of new and old has persisted. As at February 2012 the mean age of United States Air Force aircraft was: fighters 22 years, bombers 35 years and tankers 47 years.[10] The Royal [British] Air Force in 2011 retired an aircraft type which had entered service in 1965.

In 2011 the Indian Air Force was still operating an aircraft type which had entered service in 1965. In 2007 Taiwan continued to maintain a 63 year-old submarine, SS-791 *Sea Lion*. During 2011 the Indian aircraft carrier INS [Indian Naval Ship] *Viraat* commemorated 50 years in service.

Outer space reflects the terrestrial experience in which old and new technologies are used concurrently. There has been no outer space equivalent of pigeons functioning together with electronic communications, but satellite owners and operators have commonly attempted to maximise the lifespans of their satellites.[11] In June 2012 10 satellites in the Global Positioning System (GPS) constellation were Block IIA satellites, that version having been launched between 1989 and 1990.[12] Also in 2012 the average lifespan of a satellite in GEO was estimated to be 15 years.[13]

The *relationship between higher technology and the increasing time required* for design, production and distribution is also as old as recorded warfare. It was clearly faster to arm combatants with wooden spears than with siege engines. The time required to acquire and collocate all of the raw materials to cast a cannon must have exceeded the time required to make a siege engine predominantly from timber. The 19th century introduction, in the US, of mass-produced small-arms appeared to defy that trend. But on closer inspection the entire process had, as a consequence of efficient production and assembly, been reduced to a succession of lower-technology activities.

Aircraft production illustrates the relationship between higher technology and increasing timescales. During the 1930s US aircraft required a mean time-lapse of five years from concept to production. During and immediately following the impetus of the Second World War this was shortened to between three and five years. However, by the 1990s the conception-to-production time for a modified version of an *existing* military aircraft, the F/A-18 E/F, took seven years.[14] A contract for development of a subsequent military aircraft, the US Joint Strike Fighter, was let on 26 October 2001; the aircraft was not in production as at February 2014.

A similar correlation applies to submarines. During the Second World War a total of 19 German shipyards could produce between 24 and 26 U boats per month. The estimated 2012 production time for one US Virginia Class submarine was between five and six years.

Space has not been immune. The replacement of old spacecraft with newer versions involves not only innovative science but also protracted design and manufacturing time, as has occurred with technologies designed for the terrestrial environments. The time to develop space-launch vehicles, most of which are evolved versions of earlier models, appears to be in the vicinity of 10 years. The interval between concept, design, prototype construction, terrestrial testing and launch into outer space for satellites carrying highest-technology sensors can vary from four to 20 years.

The lead times for design, testing and distribution of strategic weapons and related space systems has now lengthened to the point where technologies are most unlikely to keep pace with developing strategic circumstances. In many instances their production lead-times are already longer than the duration of any war fought in the 20th century. Higher technologies in the future seem likely to further expand the design-to-production timelines.

Throughout history military commanders have taken calculated risks through *employment of immature capabilities* to gain advantages in warfare. The military vessels used by Xerxes and Themistocles were built in ignorance of the formula for displacement, which was unknown prior to c.250 BCE. Later, medieval-era cannon often exploded thus injuring or killing both operators and bystanders. The loss of a submarine in June 1774 was apparently attributable to ignorance of the rate at which water pressure increases with depth.[15] Since then, other submarines have been lost during peacetime with crew members killed in the process.

Aircraft accidents began early in the history of powered flight. The first US military death involving a motor-driven aircraft occurred in 1908 with the first such British death following two years later. Accidents during training in the First World War accounted for more than one-third of the Australian Flying Corps pilot fatalities.

Human utilisation of outer space has followed a similar path. There were to the most recent known fatality, which occurred in 2006, at least 18 deaths in spacecraft, at least 11 deaths during space-related training and a further at least 70 deaths of personnel on launch pads.[16] Between 2009 and 2011 following more than 50 years of space launch experience, launch failures continued to account for approximately 8 per cent of all launches attempted, those failures not resulting in any known human deaths.[17]

Throughout recorded history ambitious and aggressive *States have extended and then overextended their power and influence* beyond national borders. Phoenicians influenced, and at times governed, the northern coast of Africa and regions across the Mediterranean. The Roman Empire in the late first century CE included most of the Middle East, northern Africa, Western Europe and England. Religious crusades of the period 1005 – 1395 CE expanded European influence into the Middle East. The 13th century Mongol empire covered most of modern China, Russia and Eastern Europe. Subsequent colonial empires of European States included territories on the opposite side of Earth from Europe. In the 20th century Japan attempted to expand by invading Korea and China, areas of Southeast Asia and of the Pacific islands. Germany, on two occasions in the 20th century, began a program of territorial expansion across Europe.

History records that all of the ancient and renaissance empires atrophied. The later European trading and colonial empires were destabilised by antagonism between European powers and by rising indigenous independence movements. Their external territories were finally eliminated or fundamentally curtailed in the aftermath of the First and Second World Wars. When Japan and Germany attempted aggressively to expand their national territories during the 20th century they were defeated militarily.

It is almost certain extensive strategic power gained by any contemporary State will not endure. Contemporary destabilising factors include depletion of finite mineral and energy resources spurred by an addiction to endless economic growth, climate change, global population growth, the effects of increasing public and private debt, economic competition from emerging and industrialising States, and other more localised distractions such as ageing urban infrastructures. If history is a guide, the unknown issue is *when* the now powerful States will decline, not *if* they will.

The dilution of powerful-State influence is evident in outer space. Initially the US and USSR were the only States with assets in space and they remain significantly influential. Whether that bipolar dominance motivated, or simply enabled, development of similar and competitive space technologies by other States might never be known with certainty. However it is obvious similar capabilities have since been acquired by other States; the PRC, France, Japan, Israel and India in particular, but on a smaller scale.

Similarly, US global influence over a range of critical infrastructures, enabled in part by ownership and control of the NAVSTAR GPS constellation, is being challenged through the development of similar systems by Russia, the PRC, Japan, France, India and the European Union. Other competitiveness is increasing as States seek access the finite radio spectrum needed to enable space-related and many other initiatives.[18]

Throughout history new *technologies have rarely if ever been confined indefinitely to any one State*; they either proliferated or a means of negating them was invented. The ancient Romans obtained a grounded Carthaginian ship and copied it to create their first version of a quinquereme. Gunpowder, leading eventually to manufacture of torpedoes, mines and airborne anti-shipping weapons able to limit the freedom of maritime operations became available to many States. Tactically advantageous technologies including rifled gun barrels, rapid-fire weapons, turbine-powered aircraft, night-vision gun-sights and broadband communications have become commonplace. There is now almost global knowledge of the principles enabling construction of nuclear weapons.

Proliferation of sensitive technologies has affected outer space. Space-related technologies have migrated and the associated knowledge, together with geostrategic power in relation to outer space, is now distributed across three tiers. These are: the most prominent States, other spacefaring States with a technological capacity to disrupt the outer space region, and client States which have minimal influence over the outer space region. A recent and prominent example of proliferation relevant to outer space occurred when the DPRK acquired nuclear weapons as well as a space-launch or long-range missile capability.

Warfare has had an increasingly close *association with environmental pollution*, leading to possibly profound lessons for the present era. Battles of centuries past commonly resulted in regions being littered with military detritus either available for scavenging or subject to natural deterioration without enduring consequences. Following the invention of explosive munitions, battlefields could remain hazardous for decades. In one of many incidents, during August 2009 a sea-mine from the Second World War floated into a Black Sea swimming location.[19] On 3 January 2014 a bomb from the Second World War exploded in Germany killing one person and wounding 13 others.[20]

The use of nuclear weapons against Hiroshima and Nagasaki, in August 1945, resulted in pollution from radionuclides including uranium-235, uranium-238 and plutonium-239. Those events preceded the Cold War during which extensive testing was perceived as necessary to develop more powerful and efficient nuclear weapons. By 1963 States had conducted more than 500 nuclear tests in the atmosphere and by 2005 a further 1 400 such tests had been carried out beneath the Earth's surface leading to some leakage of radioactive materials.[21]

In many cases these tests and related activities were conducted by States on their own territories. In December 2011 Britain's Ministry of Defence confirmed the existence of '*at least 15 sites*' contaminated by radioactivity from activities conducted during the Second World War. The US Environmental Protection Agency in a report updated to 14 August 2012, revealed 1 000 US locations had been contaminated by radioactive materials. Military activity was not associated with all such sites but had been sufficiently prominent to justify intervention by the Defense Environmental Restoration Program. Moderate to severe radioactive contamination has been detected at over 100 sites across the former USSR. Radioactive pollution caused by military activities in the PRC has apparently become a problem, as evidenced by the People's Liberation Army issuing regulations in June 2012 to control its own radioactive waste.

One common argument implying States might not initiate warfare in outer space is there would be potential for debris to create pollution at a density sufficient to render the outer space region unusable for a protracted period. This deterrence claim is challengeable given the many terrestrial precedents of enduring pollution from explosive and radioactive materials. These precedents point to a preparedness by some States to accept enduring pollution if is considered justified by the pursuit of military or other national security objectives. An emerging and radical State, perceiving itself to be not critically dependent on space data or services, might not be deterred by the prospect of its activities adding to the density of orbiting debris.

Given the considerations to date there is obviously justification for the view that warfare in outer space is inevitable or almost so. However, other factors remain to be considered. There has, for example, been increasing awareness that constraints and limitations on human-to-human violence are desirable. Given the nexus between activities on Earth and in outer space it is reasonable to anticipate limitations on terrestrial violence might affect, to some unknown extent, attitudes toward the extraterrestrial region.

That said, the notion of voluntary restraint is complicated in all of its manifestations but seldom more so than when applied to outer space. Human activities could be affected profoundly in consequence of space warfare but the linkage might not always be immediately apparent. Most international arrangements aimed at limiting terrestrial violence are intended to protect *people*, whereas warfare in outer space would be essentially a technology v. technology battle. It is nevertheless useful to survey limitations on human-to-human violence and to consider whether even these constraints, with which it is relatively easy to empathise, have been effective.

For an example, nuclear-armed States have demonstrably resisted many temptations to use their weapons of mass destruction. Among conflicts and confrontations involving the US were: the Berlin Blockade of 1948, the Korean War, the Taiwan Crisis of 1954, the Vietnam War, the Cuban Missile Crisis of 1962, the first Gulf War 1990-91, the second Gulf War 2002-2010 and the War in Afghanistan 2001 - 2014. Argentina and Britain were at war in 1982; the USSR invaded Afghanistan in 1979 but was forced to withdraw; France fought in Algeria from 1954 – 62; India and Pakistan as well as India and the PRC have engaged in numerous border disputes. Not one of those conflicts involved nuclear weapons although in every case one or both sides could have accessed them.

Some other States have acquired nuclear weapons or embarked on development of them only to surrender their stocks or abandon the related programs. There have also been agreed and proposed several global arrangements of treaty status as well as a number of bilateral and regional agreements all aimed at limiting nuclear weapons. Those arrangements and the examples of forbearance just mentioned might support the impression of a decline in the potential for nuclear warfare. But contrary indicators also exist.

In 1949 only the US and USSR had nuclear weapons. As at 2010, known inventories of nuclear weapons were held by the US, Russia, Britain, France, PRC, India, Pakistan, DPRK and Israel. The technologies needed to build nuclear weapons had become well known. A large number of States now has access to unrefined radioactive materials. Some, possibly most, of these States could develop nuclear weapons.

Pressure applied by the international community to forego nuclear weapons has been inconsistent and selective. The DPRK and Iran have been subjected to criticism and sanctions while Israel's, India's and Pakistan's nuclear weapons have in recent decades been commonly ignored. This uneven response might encourage other States to consider acquiring nuclear arsenals either because of an impression their actions would not attract strong opposition or because they felt able to resist punitive measures likely to be temporarily imposed.

Biological weapons have a longer history than do nuclear devices. Human corpses were used to poison water wells in 1155 CE. In 1650 CE attempts were made to infect enemies with the saliva from rabid dogs. In the modern era a prohibition on poisonous gases was extended to include bacteriological weapons and entered into force in 1928. Subsequently, the *Convention on the Prohibition of the Development, Production and Stockpiling of Bacteriological (Biological) and Toxin Weapons and their Destruction* (Biological Weapons Convention, 1972), entered into force. But several attacks using biological weapons have since been recorded.

The 1993 *Convention on the Prohibition of the Development, Production, Stockpiling and Use of Chemical Weapons and on their Destruction* has received widespread international support. However, it is the most recent in a series of fragile or failed attempts to limit or prohibit such weapons. The *Strasbourg Agreement* of 1675 banned poisonous bullets in combat between France and Germany but was not further expanded. The *Project of an International Declaration Concerning the Laws and Customs of Wars* (Brussels Convention 1874) and the *Convention with Respect to the Laws and Customs of War on Land* and its Annex: *Regulations Concerning the Laws and Customs of War on Land,* (Hague II, Declaration II) each contained provisions aimed at constraining the use of chemical weapons. Chemical weapons were nevertheless employed extensively in the First World War. Subsequently, another international agreement was drafted prohibiting their use (the *Geneva Protocol for the Use of Asphyxiating, Poisonous or Other Gases, and Bacteriological Methods of Warfare* (Geneva Protocol, 1925)). However several States began stockpiling chemical weapons, indicating an intention to ignore the Protocol and to use the weapons if considered necessary. More recently, Iraq used chemical weapons between 1981 and 1988. Chemical weapons were released in Syria during August 2013.

Incendiary devices were used by the ancient Egyptians and by other States subsequently. This led, in 1139 CE, to the Second Lateran Council's ban on '*incendiarism*'. Fire has nevertheless been employed in warfare ever since with napalm having been used by at least 14 States since 1940. At present, international law restrains the use of incendiary weapons against civilians but not against military personnel. (*Convention on Prohibitions or Restrictions on the Use of Certain Conventional Weapons which may be Deemed to be Excessively Injurious or to have Indiscriminate Effects, Protocol III, Protocol on Prohibitions or Restrictions on the Use of Incendiary Weapons* (Incendiary Weapons Convention, 1980).

Landmines have been employed since non-explosive spikes and stakes were set by armies in 500 BCE and possibly beforehand. The earliest description I found of a pressure-operated landmine was dated 1726 CE. The current (1997) Ottawa Treaty, *Convention on the Prohibition, of the Use, Stockpiling, Production and Transfer of Anti-Personnel Mines and on their Destruction,* prohibits landmines. It has the support of 161 States parties but States not signed include the US, PRC and Russia. The *Landmine Monitor* for 2013 recorded 12 producers of antipersonnel mines while terrorist organisations, together with three States, were accused of actively laying land mines. A total of 3 628 new casualties had resulted from land mines.[22]

Cluster munitions were used by German and USSR forces during the Second World War. Subsequently, 36 States were claimed to have produced them and 18 to have used them.[23] In May 2008 a Convention was adopted banning cluster munitions, *Convention on Cluster Munitions.* The US, PRC, Russia, India, Israel and Pakistan were among States which had not signed the Convention by January 2014. After the Convention took effect cluster munitions were used by Thailand and by Libya.

Some other less predictable aspects of warfare also challenge, but do not eliminate, the impression of an inescapable trend toward greater lethality enabled by technology.

Military forces do not invariably recognise or apply the latest technologies, although to be fair they often do so. Evidence of crossbows was found in Chinese tombs dated as 4th century BCE but the weapon appeared in Europe 1 700 years later and then might have been an indigenous invention. Large cannon could not at first be used aboard ships because the recoil from firing would damage or wreck the vessels. It took 200 years to identify the technique of lashing cannon with ropes to stretch and absorb recoil forces. The technique of 'rifling' gun barrels, resulting in improved projectile accuracy, was known in Europe by c.1476 but gained acceptance gradually and was still being adopted incrementally during the 19th century.

Instances of reluctance to break with entrenched behaviours or valued traditions have obscured changes to tactical and strategic circumstances. The Middle-Eastern Mamelukes abhorred firearms, preferring to rely on their skills with horses and sabres. They were defeated when confronting the muskets of the Ottoman army at the battle of Marj Dabiq in August 1515 and again at Raydiniya in January 1516. The 16th century Japanese samurai shared the Mamelukes' revulsion for firearms and effectively isolated Japan from them. Japan became strategically weakened, a circumstance revealed in 1853 when an American, Commodore Matthew Perry, visited Japan. In August 1942 a squadron of Italian cavalry performed its final charge by attacking a Soviet infantry regiment.

Compulsive traditions have also influenced weapons and related accessories. Medieval body armour gained in weight to the point where some metal-clad knights of the 14th and 15th centuries could no longer fight on horseback. In the 15th century the Turks began using cannon to batter the walls of Theodosius at Constantinople. Eventually their cannon became so large and heavy the weapons had to be cast *in situ* and presumably in view of defending forces.

During the Second World War Germany built the prototype of a battle tank, the *Panzerkampfwagen VIII*, with a mass of 191 tonnes, making it too heavy to drive over most bridges. In 1983 the mass of US F/A-18 aircraft, navy version, was so great as to preclude take-off from some US aircraft carriers unless fuel or ordinance were first reduced.

History also records many incidents in which stalemates, truces, retreats or outright defeats eventuated despite the losing side having superior technologies or larger numbers of equivalent technologies. These outcomes point to the influences of leadership, morale and related factors which at times can be sufficiently potent to modify or even negate benefits conferred by technology.

The US-led military engagement in Vietnam (c.1964 – 1975) ended in an American withdrawal despite the US having had a technologically superior force and dominance of the air environment. The Six Day War (1967) involved Israel on one side confronting Egypt with the support of Arab nations across the Middle East. Israel captured approximately

119 140 square kilometres of territory, suffering fatal casualties at an Israeli to Arab ratio of 1:6 and destroying opposing aircraft at a comparative loss ratio of 1:10. The USSR intervened in Afghanistan beginning in 1979. When it withdrew in 1989 the USSR's superior technology, military professionalism and mobility had achieved little. The USSR intervention followed a history of failed attempts to subdue tribal Afghanistan, including the First Afghanistan War (1838 – 1842).

Summarising, except for the Antarctic continent every terrestrial landmass has experienced bloodshed caused by human-to-human violence. In near-Earth outer space, the Moon appears to have remained free from overt aggression and from space weapons tests. These precedents could indicate some potential for outer space to become a zone of peace, provided contrary evidence is ignored.

Antarctica is more complicated and less geopolitically stable than might be implied by the absence of overt warfare. I have considered this further in Chapter 6. Concerning the Moon as an extraterrestrial haven of peace, in 1959 the US planned to install a military base on the Moon by late 1966 (Project Horizon), although the project did not proceed. Other attempts or proposals to influence or restrict access to lunar real-estate are also mentioned in Chapter 6.

This highly compressed history of warfare illustrates the existence of a well-established quest for increased range and lethality of weapons. Coincident with expansion of direct and indirect military competencies, there has been a trend toward greater complexity of weapons associated with longer timelines for their development and availability. Former distinctions between military operations on or in the land, sea, sub-surface, air and outer space environments have become blurred. All of these influences, and other with long terrestrial histories, are evident in outer space. However, the combination of voluntary constraint by powerful States, attempts to limit the use of some categories of weapons, and the unpredictable components of warfare combine to render predictions regarding the future course of warfare, either terrestrial or in outer space, potentially hazardous.

Nevertheless, the threat of war *per se* is unlikely to disappear. Summarising the most recent 5 000 years of warfare, globally, David Chandler, former head of War Studies at Britain's Royal Military Academy, Sandhurst, has written, *'Over the past five millennia there have been only 290 years free of the horrors of organised armed strife (not including guerrilla activities).'*[24]

Postscript

As a postscript, the following two diagrams illustrate the most common satellite orbits. They provide an orientation to information in chapters following.

Figure 1 – 1. Orbits of Artificial Satellites Tending to Equatorial Orientation [25]

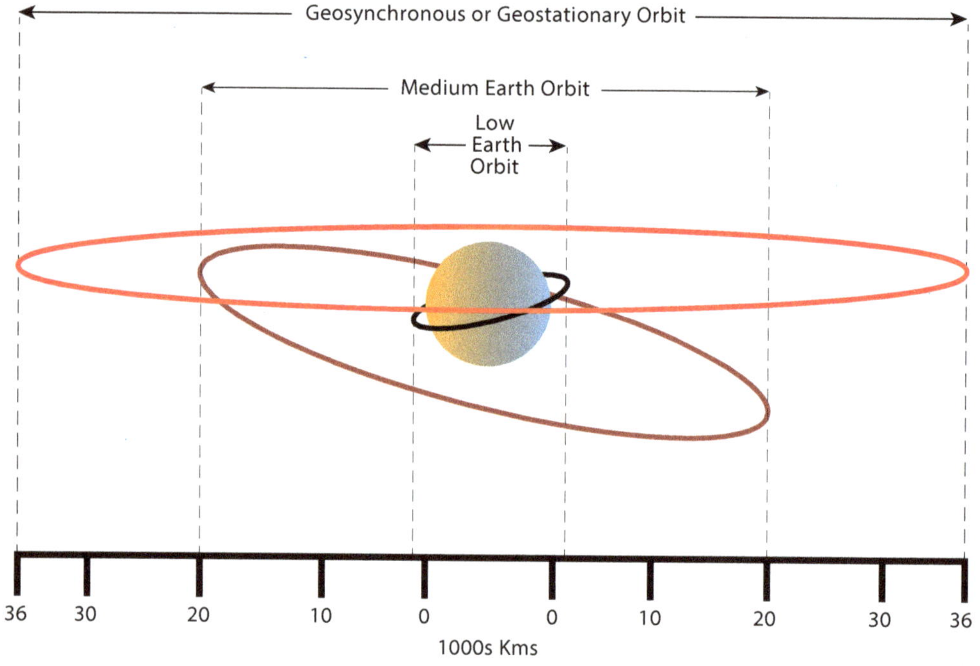

Notes:

1. There are no formal definitions of the dimensions of orbits for artificial (viz. human-made) satellites. Those shown represent commonly accepted concepts and applications.

2. Many, but not all, orbits are inclined relative to either the equator or to the North and South Poles. For example, the International Space Station orbits with an incline of ~52° relative to the equator. If the orbit of every active satellite was shown, the illustration would resemble a mesh covering Earth with lines at many angles but also with some concentrations of lines at specific angles.

3. Terminology associated with the orbits of artificial satellites includes: *geocentric*, orbiting around planet Earth; *apogee*, the highest altitude achieved during an orbit; and *perigee*, the lowest altitude achieved during an orbit. Deductively, most orbits are elliptical to some extent.

Figure 1 – 2. Orbits of Artificial Satellites Tending to Polar Orientation [26]

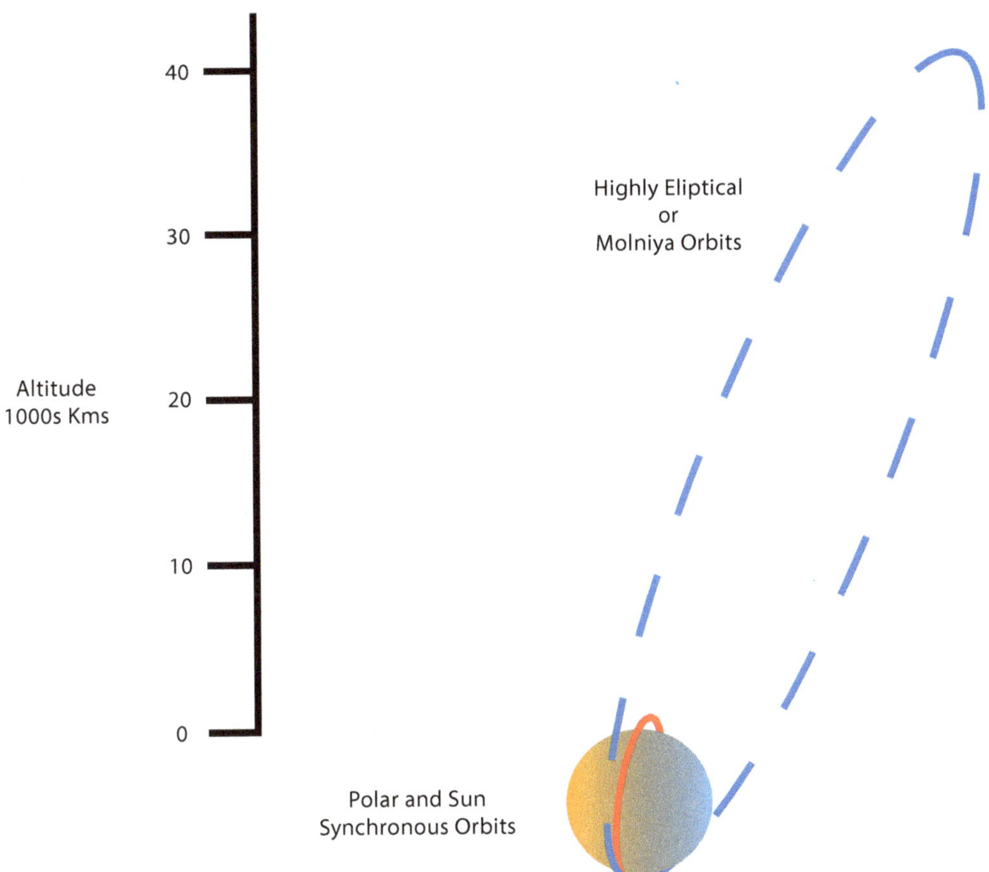

Notes:

1. The low orbits with polar orientations are broadly characterised as polar or sun synchronous. Satellites in these orbits tend to travel at approximately constant velocity. Satellites in the highly elliptical orbits, sometimes referred to as Molniya orbits, travel more slowly as altitude increases. They accelerate as they approach Earth on each orbit.

2. In this figure the highly elliptical, or Molniya, orbit is shown in the northern hemisphere. The principles governing geocentric orbits also allow for satellites to be launched into such orbits with a southern emphasis. Thus far (2014) demand has been insufficient to attract the necessary capital and other support for launching a satellite into the southern profile.

Chapter 2. Some Constraints Affecting Outer Space Weapons

Space weapons have a long pedigree relative to the duration of humankind's experience with artificial satellites. On 16 October 1957, only 12 days after *Sputnik 1*'s launch and before the US had placed any satellite into orbit, the US Air Force (USAF) fired pellets into '*orbital and escape velocities*' at an altitude of 87 km (the Aerobee test).[1] This might have been the first launch of the anti-satellite weapon later to be described as a space-mine.

These and other so-called space weapons are commonly components of systems. The systems include a decision to mount an attack, availability of sensors to detect and track a target spacecraft, capacity to deliver a weapon to within effective striking distance of its target, a technology intended to terminate or curtail the target's operation, desirably a means of evaluating the success or otherwise of an attack, and a command system to coordinate the entire process. In this work I assumed weapons able to attack spacecraft would have been enabled by a conscious decision to attack and that the process from decision to execution would have been systemically managed, even if the actual weapon was able to function autonomously in the advanced stages of target engagement.

Information and data concerning space weapons up to the mid 1990s is now commonly accessible. More recent material is less easily located and less detailed. This implies, not unexpectedly or unreasonably, most projects associated with space weapons are classified for security purposes.

I therefore obtained information regarding more recent space weapon developments from a variety of publically available sources many of which lacked the imprimatur of a State agency, although official sanction does not always guarantee accuracy. I also interpreted available data cautiously and consulted a wide range of independent and concurrently published texts. When published data appeared inconsistent with fundamental science, I either mentioned the inconsistency or discarded the data.

I have treated ICBMs, intended primarily for Earth-to-Earth warfare, as having a secondary role as outer space weapons. Trajectories of some of these missiles take them to altitudes higher than the lowest limit of outer space, accepted here as 80 km, and also higher than the altitudes of many orbiting satellites.[2] The missiles could be detonated in outer space, resulting in the destruction of satellites. It is also plausible that, '*... much of the current US development of space-based technology and weaponry, including space-based interceptors and airborne and space-based lasers, is taking place under the guise of missile defense.*'[3]

Space weapons can be categorised in relation to the targets they can attack, in accordance with the destructive technology they employ, or by the way they are moved into a position from which they can attack. This investigation gave primacy to the destructive technology employed, although potential targets and means by which the weapons are manoeuvred into position for an attack were also identified where relevant.

However, categorisations of outer-space weapons in accordance with the destructive technology employed can be confusing as well as irrational. The commonly employed terms: 'nuclear', 'directed energy' and 'kinetic' each refer to a category of space weapon. But nuclear is effectively an energy source, while directed energy is a category of techniques for transmitting energy toward a target and kinetic refers to a class of energy. Nuclear and directed energy weapons can exert kinetic effects. I nevertheless reluctantly retained those three categories, among others, because they reflected common usage in literature related to space weapons and because employment of unique terminology could cause yet more confusion.

Impressive claims abound regarding weapons designed to be used in outer space, but enthusiastic advocacy often brushes lightly over four key issues. How much power, convertible to energy, is needed to damage or destroy spacecraft? What can supply that power? Can sufficient energy be applied to cause the necessary level of harm within an available window of opportunity? And, for weapons intended to reside in outer space, can their energy sources be launched into space and maintained there? The questions are simple enough but acceptably precise answers are elusive.

One incontrovertible constraint is that weapons in outer space, as on Earth, must comply with the first law of thermodynamics; energy cannot be created or destroyed, it can only change form.

Power/Energy Required to Damage Spacecraft

Exposure to a relatively minor level of destructive energy could catastrophically damage some spacecraft if the energy was applied over a weeks or months, and accumulated by the target, until it reached a critical level. This might occur when, for example, spacecraft encounter, during some part of each orbit, the residual effects from a nuclear detonation in outer space. Appendix 2, *Nuclear-Detonation-Induced Radiation Effects* (NDRE), contains more detail. A contrasting situation would arise when a much greater concentration of energy was needed to take advantage of a fleeting opportunity to eliminate an apparently hostile spacecraft transporting a nuclear weapon.

Most potentially aggressive encounters with spacecraft might not mirror either of these extreme examples and this is where areas of doubt and confusion start to emerge.

Consider the destruction of spacecraft by applying melting or ablating for example. Estimates of energy/time required to achieve the desired intervention require information on factors for which accurate data might not be, in fact almost certainly will not be, available. Both before and during the 21st century many commercial satellites were sheathed predominantly with the polyimide film Kapton which remains stable to approximately +400°C.[4] However, alloys designed for high-stress terrestrial applications can remain stable to 980°C or higher. States concerned to protect their most strategically valuable spacecraft from attack could be expected to suppress data regarding the thermal properties of sheaths or additional protective shields.

Additionally, spacecraft can utilise screening and insulating technologies to dilute the effects of heating or ablating energy to which they are exposed. Some of these capabilities can be accessed by reviewing the security-unclassified specifications of non-military and other scientific spacecraft.

The surface of a reflective sunshade attached to the spacecraft *Messenger* which began orbiting Mercury in 2011, was expected to reach ~ 370°C. However, *Messenger's* insulating technologies were designed to provide an operating environment of ~20°C for thermally shielded components of the spacecraft. The leading edges of wings on the former US Space Shuttles could withstand up to 1 260°C during re-entry into the dense atmosphere while limiting the temperature of the spacecraft skin to ~177°C. An aggressor intending to rely on melting or ablating effects could thus infer some approximate specifications regarding protective technologies employed on a targeted spacecraft, but the margin of error might be significant.

A further variable will be the temperature of a spacecraft's sheath at the moment preceding an attack. This will normally affect the amount of applied energy needed to cause melting. Objects in Earth-centric orbits are subject to several environmental (thermal-related) influences.[5] Thermal energy generated by operating components of the spacecraft may also be evident on its external sheath. A related variable will be the capacity of a target spacecraft to reduce accumulated heat by radiation, as many space vehicles are designed to do.

Adding yet more uncertainty to calculating this destructive energy threshold is that spacecraft *components* can operate over a range of thermal conditions. Moreover the rate at which heat can be transferred to them from the spacecraft's sheath will be unknown unless details of the target's design and related materials as well as structural engineering specifications are available. This issue becomes particularly relevant if damage to a spacecraft's sheath cannot be achieved and reliance is then placed on the transfer of energy from the sheath to internal components, by either radiation or conduction depending on spacecraft design.

Despite these challenges some estimates of necessary power/energy levels associated with melting or ablation have been attempted, although the basis for data published was either incomplete or obscure. Aftergood, 1985, estimated the average power requirements impliedly to cause catastrophic damage as, for a space-based chemical laser weapon, 100 – 200 kW, and for a '*neutral particle beam/space-based free electron laser*' (FEL) weapon as being 1×10^5 kW – 5×10^5 kW.[6] Given the wide discrepancy between estimated power for the chemical laser and the beam, or FEL, this data deductively applied to total input power.

The efficiency with which input power is converted into the energy transported to its target is further considered in Chapter 4. However, indicatively the total input to optical output efficiency for a chemical laser, illustratively a chemical oxygen-iodine laser (COIL), is ~ 32% and for a FEL, ~ 24%. Using those conversion factors and the above estimates of required power, the COIL would apply ~32 – 64 kJ/s to its target whereas the FEL would apply 24 000 – 120 000 kJ/s which, assuming identical targets, does not make sense. A partial explanation for the discrepancy possibly lies with fundamentally distinctive input/output conversion efficiency calculations used for lasers and other technologies but they were not disclosed.

Aftergood did not specify the area of target covered by the projected concentration of energy. If, for example, to concentrate an average of 24 000 kJ/s on 1.0 cm^2 required 10^5 kW of power, then a circular beam with a radius of only 3.0 cm, covering an area of 28.27 cm^2 ($A = \pi r^2$), assuming a plane target-surface transverse to the beam, would absorb 28.27×10^5 kW to apply the same uniform concentration of energy.

A separate calculation by the US Office of Technology Assessment was that thermal energy necessary to disable space vehicles would be in the range '*1 – 100 kJ/cm²*'. The associated text stated, '*This energy must be delivered quickly – if the time taken to deliver a lethal amount of energy is very long ... the heated area ... may have time to conduct away much of the energy being directed at it and may not fail.*'[7]

The broad range of that estimate pointed deductively to recognition that the time a weapon would be required to dwell on its target, and other relevant factors, might not be known, as I have indicated.

Taking account of factors considered, and those identified above were not exhaustive, precision is obviously not possible. Nevertheless, an indicative arbiter seems desirable to allow for at least a generalised estimate of whether a claimed space weapon intended to achieve melting or ablation might be practicable. The figure I have accepted is the mid point of the Office of Technology Assessment's range, namely *50 kJ/cm²/s*. This is an unsubstantiated decision, but no conflicting estimates were found nor seemed likely to be found. Nevertheless, it is not an inflexible criterion, merely an indicative yardstick.

Regarding damage by radiation, in 1986 US researchers hypothesised that an absorbed dose of 10 k grays (Gy) would permanently damage most radiation-resistant, high-density, silicon-integrated circuits although in some instances a significantly lower dosage would suffice. [8] The 10 kGy estimate appears to have remained valid as a destructive radiation threshold.

Components on the Mars explorer, *Curiosity*, were reported as hardened to 1 kGy. Reference was found to a BAE Systems Electronic Solutions product, the RAD750 single-board computer, being hardened to 10 kGy. I did not locate higher ratings for any space-intended component or application. I note also that resistance to radiation by older satellites is likely to have been reduced to levels *below* their original component ratings because they will have already been exposed to accumulated radiation from natural sources. Given the information and data available, I accepted 10 kGy as the level of absorbed radiation able to damage most spacecraft.

I additionally accepted the yields of isotropic nuclear weapons as adequate to inflict catastrophic damage on spacecraft. This was notwithstanding the variations in yield between weapons based on either fission or fusion physics and regardless of spacecraft design. The energy produced by a nuclear detonation will be approximately 50 % blast energy, 15 % nuclear radiation and 35 % thermal energy. A 1.0 Mt nuclear weapon when detonated will produce ~4.18×10^{15} J.

Primary variables governing the capacity of a nuclear weapon to almost immediately and catastrophically damage spacecraft are the energy released by the detonation and the linear distance between the detonation and a target spacecraft. Given terrestrial precedents there is no doubt the detonation of a 1.0 Mt weapon within a range of 1.0 km from a spacecraft will result in its destruction, although even more spectacular outcomes are also possible and will be considered.

The formula governing kinetic energy on rigid bodies ($E_k = ½ mv^2$) is well known. However, many variables could affect the capacity of a kinetic-energy device to damage or destroy a spacecraft. These include: whether the target is approaching or receding relative to a given projectile, the physical shape of the target spacecraft, the angle at which energy is applied, the physical composition of the target, the target's engineered capacity for self-protection, and whether the attack involves one projectile or many of them.

Nevertheless, in considering the destructive potential of kinetic weapons it is appropriate to note the kinetic energy of a solid object travelling at closing speed equal to that of a spacecraft in low Earth orbit (~ 7.0 km/s) is far in excess of the energy available from an equivalent mass of 2,4,6-trinitrotoluene (TNT). In the 1960s a study undertaken in the US calculated an effective warhead could be built weighing '*as little as 2 lbs* [0.907 kg].'[9] Allowing for a >500 % enhancement to the protective sheaths of 1960s spacecraft, a kinetic warhead of ~ 5.0 kg mass impacting at orbital velocity could cause catastrophic damage and 5.0 kg is an insignificant payload mass in contemporary terms.

Precise specifications of contemporary kinetic warheads were not located although my search disclosed one design for a missile warhead and missile structure to be combined into a single entity, the implications of the mass of any such configuration being obvious. A reference to energy delivered by the warhead of a US SM-3 missile described it, believably, as the equivalent force released '*when a ten ton truck travelling at 600 miles per hour* [965 km/h] *hits a wall.*' [10]

In summary, the combination of innate variables implies it will be almost impossible to accurately determine the kinetic energy required to damage or destroy a space vehicle, except by modelling which will be unlikely to reflect the combination of factors, weighted appropriately, of a real encounter. However, as will be considered, kinetic technologies have been among the most successful of weapons used in outer space. It is deductively clear these weapons were designed to surmount all known forms of impact resistance by many, and possibly all, spacecraft provided closing velocity is adequate. Consequently, the minimum quantum of kinetic energy required, although not publically known, and possibly not officially known either, is innate to these weapons. They can demonstrably deliver it in outer space with a very high likelihood of success.

Sources of Power Required

The following discussion examines potential sources of power to enable weapons in outer space. Clearly, if adequate power is not available then a proposed weapon will be unable to deliver enough destructive energy, regardless of how efficient or lethal the theoretical engagement might be. Nuclear reactors, solar-energy collectors and fuel cells are considered. Discussion of radioactive sources dominates because the related technologies have been already been used successfully to generate high levels of power / energy in outer space.

During the 1960s and 1970s, the power requirements of spacecraft were high compared against contemporary requirements. Technologies to collect and store energy in space were then less efficient than at present.[11] Consequent limitations constrained the duration of space missions, limited the instrumentation able to be placed into orbit, and led to consideration of employing nuclear reactors as sources of electrical energy for Earth-centric spacecraft.

By the final decades of the 20th century, solar arrays could meet the modest power requirements of satellites and other spacecraft in Earth-centric orbits. Deductively, the remaining reason for positioning a nuclear reactor in any such orbit would be to enable a space weapon, unless unanticipated and power-hungry instruments were launched.

I nevertheless have included the following synopsis of nuclear reactors designed to be used in outer space. It remains relevant because it identifies States which have the capacity, and have in the past shown political will, to employ nuclear reactors in space for any reason, including possibly to enable a space weapon. This review does not extend to radioisotope thermoelectric generators. These and other lower-level radioactive power sources, typically producing ≤ 1 kW, have no apparent application in supporting space weapons except possibly to enable devices such as dazzling lasers in which role nuclear power is but one option among others.[12]

The US appears to have been the first State to launch a nuclear reactor into outer space. That event, in April 1965, was one of a series of experiments conducted under the heading, System for Nuclear Auxiliary Power (SNAP). The launched reactor, *SNAP 10A,* was fuelled by 93 % enriched uranium-235. It could provide electrical power in the range 30 – 45 kW. The *SNAP-10A* had a mass of 454 kg.

A subsequent version, the *SNAP-8,* which was never launched into outer space, could achieve 600 kW but had a mass of 4 545 kg without shielding. During 1986 another reactor,

SP-100, was designed to produce 100 kW of electrical power. Its mass was ~ 5 422 kg including 1 255 kg of shielding, although several possible variations in design specifications would have both increased or decreased total mass. The *SP-100* reactor design was intended to be scalable.[13]

Recalling my approximations for ablating and melting, and using an electrical power to transmitted energy conversion factor of 24 % (nominal FEL), the projection of 50 kJ/cm2/s would require input power of ~ 208 kW per cm^2; thus the *SNAP 8* could nominally have supported a FEL beam covering a plane area, transverse to the beam, of ~ 3.0 cm^2 at that energy level. The *SP-100* configured for an output of 100 kW would have enabled the application of 50 kJ/cm^2/s to an ~ 0.5 cm^2 surface.

(To keep this conversation appropriately broad and readily intelligible, issues such as beam diffraction, collimation, divergence and distance between sensors and targets, although clearly relevant, were not considered in these and following examples. Had they been included, the potential requirements for power would have increased.)

In either case cited, some damage could conceivably have been inflicted on a spacecraft by applying less energy per cm^2, but the impression given is not one of power sufficient to enable a potent outer space weapon.

According to the World Nuclear Association, in the 1980s US Department of Defense developed, at least to the conceptual stage, another nuclear reactor at a scale beyond the requirements of a civil space program.[14] Either the related project or the reactor itself was known as Timberwind.

Beginning in the early 1970s, observations made by the US implied the USSR was using nuclear reactors to provide electrical power for satellites. This was confirmed in 1978 when the USSR satellite *Cosmos 954*, carrying a nuclear reactor, re-entered the dense atmosphere.[15] The USSR subsequently developed another nuclear reactor for use in space, the Topaz-2, launching two of them into orbits of approximately 800 km altitude on *Cosmos 1818* and *Cosmos 1867* in 1987-88. The reactors performed at '*a power level of 10 kilowatts electric*'. They each were attached to 7.0 m^2 radiators. Topaz was subsequently considered for use on a joint US-Russia program, possibly involving Britain, but funding was withdrawn before details of the program could be finalised.

So far as I could determine, not one of the USSR's nuclear reactors placed into orbit could have powered more than satellite sensors or interplanetary exploration vehicles. No information located has implied that the USSR's reactors were scalable or that research aimed at producing a nuclear power source for directed-energy space weapons had been undertaken by the USSR, although it might have occurred clandestinely.

The issue of nuclear reactors with constant output insufficient to power a potent space weapon opens the possibility of using capacitors to deliver pulsed power. However, to achieve a required peak-power level, available power must first be accumulated. The greater the difference between available constant power and the peak discharge required, the longer must be the time interval between pulses emitted by a given capacitor. The proposed weapon must be compatible with a pulsed power input. The delays could provide time for a target spacecraft to shed some of an additional thermal load, particularly through radiation. Nevertheless, capacitors have a potential role in situations where the peak pulse obtainable is fed to a weapon with pulse-compatible technology and is sufficient to inflict significant and irrecoverable damage to a spacecraft.

Considering their level of technological achievement, including developments associated with using terrestrial nuclear energy, it seems probable Israel, India, Japan, Britain and France have the technology and capacity to develop a compact nuclear reactor. Given the missile and launch vehicle capacities of those States, considered subsequently, they could launch such a nuclear reactor into outer space, except for Britain which has no indigenous space-launch capability unless its long-range missiles obtained from the US were modified for that purpose. I did not find records, or hints, of those States developing nuclear reactors of a capacity adequate to power a directed energy space weapon.

The PRC, with access to powerful launch vehicles, also considered subsequently, has been developing terrestrial nuclear power generation. In the context of its other advanced technological developments, it seems likely the PRC could deploy nuclear reactors to outer space if the benefits of using this technology were identified. Former PRC President, Jiang Zemin, has recorded strong support for using nuclear energy power sources. A trend to expanding PRC use of nuclear-generated energy is established and has been forecast to continue.[16] No reference I found indicated that the PRC had undertaken research aimed at producing a nuclear power source for directed-energy space weapons.

A 2005 report released by the International Atomic Energy Agency (IAEA) listed nuclear reactors being considered for space applications. They included nuclear reactors rated as capable of 10 – 100 MW electrical. It was not clear from the publication whether all of these reactors were extant or conceptual.[17]

Internationally, research into nuclear reactors is continuing with most investigations concerning terrestrial applications. Some research activity could nevertheless support deployments in outer space. During 2008 the US was reported as investigating a new nuclear reactor intended for deep space missions. The overall project, JIMO, was associated with applications requiring a nominal power level of 200 – 2000 kW for a functional life of approximately 10 years.[18] The peak power level in that range could enable melting or ablating weapons with beam coverage of ~ 10.0 cm^2 at 50 $kJ/cm^2/s$, assuming a conversion efficiency of 24%, or a larger coverage with less energy applied to the target.

Reductions in nuclear reactor mass might become possible in part as a consequence of continuing research being conducted into innovative fuels. Research aimed at enhancing electrical conversion efficiency is also progressing with potential benefits for both terrestrial and space applications. Available data and information implied nuclear reactors suitable for orbiting in outer space and providing power at a level sufficient to enable a space weapon could conceivably exist in 2014 although I found no direct evidence to support such a claim.

Even if these devices do already exist in classified national facilities, launching radioactive materials into outer space involves unique hazards and challenges including the potential to attract adverse international reactions. If damaged during a conflict or at any other time an orbiting nuclear reactor could distribute radioactive material which, after a period as space debris, would re-enter Earth's dense atmosphere in a random and uncontrolled manner. However, in some cases re-entry could be managed or postponed by planned intervention. This could include recapture from orbit, injection into an escape trajectory, or repositioning into a high orbit (which would bequeath to future generations responsibility for dealing with associated problems).

When coupled to a space weapon the peak-energy of a nuclear reactor would be needed only occasionally, conceivably once. To conserve fuel and prolong useful orbital life the reactor would almost certainly be held close to a sub-critical configuration. Even in this mode, nuclear reactors continue to produce heat energy and, in outer space, that energy must be radiated away. Radiated emissions from a nuclear reactor in space would identify it as a potential target to be attacked in the case of conflict.[19] Its signature would be strongest and easiest to detect if the reactor was powered-up to support a weapon. A nuclear reactor in outer space coupled to indispensable radiators would also form an unconventional and therefore suspicious physical profile, a profile able to be recognised by electro-optical, radar or hyperspectral sensors and possibly also by conventional telescopes.

Orbiting nuclear reactors would also require protection from natural and artificial hazards. A spacecraft manoeuvring capacity would be necessary to reduce potential vulnerability to damage from detectable items of debris. The manoeuvring capacity would utilise fuel as would a requirement to confine the host spacecraft to its assigned orbit. The amount of fuel required for manoeuvring would increase if the host spacecraft employed rapid evasion or orbital transfer capabilities. The rate at which fuel for station-keeping or manoeuvring was consumed would be a primary determinant of on-orbit lifespan, as it is for almost all spacecraft, unless on-orbit refuelling becomes practicable.

In summary, resolution of technical challenges associated with designing, building and launching into outer space a of powerful nuclear reactor will not be a key to military success so much as the entrée to an additional set of fundamental challenges.

Solar energy, convertible to electrical power, is a consistently available source of energy in near-Earth outer space. However, a range of influences affects the efficiency of photovoltaic conversions. In particular, the intensity of total solar energy in near-Earth space varies as a consequence of the solar cycle and also because of changes reflecting cyclic variations in the distance from Sun to Earth that occur during Earth's annual orbit of the Sun. For estimates of solar energy in near-Earth space-over-time, I have adopted a mean value of 1.36 kW/m^2 for the solar constant.[20]

The best conversion rate to electrical power I could find for any category of photovoltaic cell was 43.5 % achieved in 2012. [21] However, given past progress it is probable further enhancements to conversion efficiency will be achieved as a consequence of developmental work continuing in institutions globally. Allowing for technological advances a conversion efficiency rate of 50 % (providing 0.68 kW/m^2) will be assumed as achievable in outer space by 2030. On this basis, acquisition of a modest 208 kW, enabling a hypothetical FEL to apply ~50 kJ/cm^2/s over 1.0 cm^2, would require a solar-energy collection array of (208/0.68 =) ~306 m^2.

Giving context to the dimension of that hypothesised solar array, in 2014 the US National Aeronautics and Space Administration (NASA) was building a collapsible sun shield of ~300m^2 intended for launch into outer space. Thus, on first impression the solar array appears to offer an achievable solution to the challenge of providing power sufficient to enable a directed-energy weapon in outer space. However, if an ~306 m^2 solar array could enable a FEL to project ~ 50 kJ/cm^2/s for 1.0 cm^2 then coverage of a target area of, say, ~ 50 cm^2 (a circular beam of ~ 8.0 cm diameter) with the same concentration of energy would require 10 400 kW and therefore a solar array of (10 400/0.68 =) ~ 15 294 m^2, an area slightly larger than two international soccer-football fields. Even if the true requirement was only half of that shown it would pose a significant engineering challenge, but not necessarily an unachievable one by 2030 CE.

Theoretically, a smaller solar array could collect energy for accumulation. There have been continuing advances with storage devices broadly classifiable as batteries; for example ultra-capacitor chemicals or superconducting magnetic energy storage. However, storage in outer space of the power required to enable any space weapon would present new space-launch and engineering-design challenges while bringing considerations of battery life/efficiency into the overall energy acquisition and release equation. A smaller collection array possibly coupled to an integrated storage device could power a capacitor capable of delivering a significant pulse, as already discussed. But any such architecture would incur the same limitations as previously considered if protracted or continuous applications of energy were needed to overwhelm the capacity of a target spacecraft to defend itself.

To offer maximum efficiency in acquiring solar energy the array would also need to be maintained on a plane facing the Sun. But an Earth-centric-orbiting solar-energy-collection structure would be in eclipse relative to the Sun for variable diurnal periods depending on a number of influences. These include the position of Earth along its orbit around the Sun, the altitude of the host spacecraft relative to Earth, and the angle of the host spacecraft orbit relative to Earth's geographic equator.

For a satellite orbiting in geosynchronous orbit (~ 35 780 km mean altitude) the period of total eclipse would seasonally reach a maximum diurnal of ~72 minutes although entry into and exit from the masking effect of the dense component of Earth's atmosphere would effectively extend the period in which solar energy could not be efficiently collected. The duration of this eclipse, or occultation, would be longer for satellites in most of the lower orbits.

Solar arrays are vulnerable to space weather. Solar flares and coronal mass ejections reduce collection efficiency by between 3 and 5 % during each event. Other cosmic events can permanently damage solar arrays in space. To minimise damage from collisions with meteoroids, particularly during peak meteoroid events, the operators of satellites sometimes reposition spacecraft to appear perpendicular to the anticipated angle of meteoroid approach. Solar energy collection would then be disrupted or degraded.

Considering these limitations and associated challenges it seems likely solar energy cannot yet enable a high-powered directed-energy weapon for use in outer space although such a feat might be achieved by 2030. Any such solar array would inevitably present a large and obvious target to be attacked in time of outer-space conflict.

Fuel cells might eventually become available to support high-powered weapons in outer space, and several categories of fuel cell have already been used there for other purposes.[22] NASA has investigated potential applications for proton-exchange-membrane fuel cells, regenerative fuel cell systems and solid-oxide fuel cells. US Space Shuttles carried three independently-operating fuel cells utilising hydrogen and oxygen fuel while generating water as well as electric power.[23]

However, fuel cells produce prominent and therefore easily detectable signatures in the infrared spectrum. Phosphoric acid fuel-cells commonly operate in the thermal range 150 – 200°C while converting the energy of their fuel to electricity at efficiencies in the range 37 – 42 %. Molten carbonate fuel cells can convert fuel to electrical energy at 65 % efficiency but have operating temperatures of ~ 650°C, again with implications for detection in outer space.

Fuel requirements will further erode the practicability of high-output fuel cells in outer space. Indicative fuel consumption is that a hydrogen-enabled fuel-cell would require approximately 5 550 kg of hydrogen to provide a constant 0.5 MW for 20 minutes. [24] Fuels other than hydrogen, commonly associated with fuel cells, are not likely to fundamentally reduce the mass / energy-requirements for fuel, as illustrated by the table following (accepting that mass / energy ratios are not the only criterion of fuel cell efficiency and observing that regenerative technologies might improve).

Table 2 – 1. Energy Density by Mass, Selected Fuel Chemicals Used in Fuel Cells[25 26]

Chemical Fuel Component	Energy Density by Mass (MJ/kg)
hydrogen (high heat value (HHV))	141.0
methanol	22.7
methane	55.5
ethanol	29.7
carbon (to CO_2)	32.8

The possibility that fuel cells might in the future be employed to support space-based weapons cannot be unconditionally dismissed, particularly for deployments to the lower orbits. Promising claims have been made in relation to hydrogen peroxide cells. Regenerative fuel cells could reduce reliance on initially-launched fuel, particularly if some or all energy for regeneration was provided by an integrated solar array. The power output from one or more fuel cells could be increased by integration with capacitor technology.

In summary, fuel cells remain an unlikely although not inconceivable source of future power for weapons in outer space. For fuel cells to become practicable in that role, challenges associated with fuel mass would have to be resolved. The problem of detectable infrared signatures in outer space would need to be addressed in ways not presently apparent.

Launching and Resupplying

Launch capacity will constrain the mass and physical dimensions of any power source, anti-satellite weapon, or any other item, intended for outer space. As at 2014 indicative maximum payloads for launch vehicles releasing objects into low-Earth orbits were: Atlas V 551/552, 20 520 kg; Delta IV Heavy, 23 040 kg; and Proton KM/Breeze, 20 000 kg.

More powerful launch vehicles were being developed to carry payloads of approximately double the individual maxima shown. But as a generalisation maximum payloads for launches into geosynchronous orbit are typically almost half of that able to be positioned in a low orbit. This limitation is unlikely to change.

The dimensions of US Space-Shuttle cargo bays, when those vehicles were operating, were ~18.3m length and 4.6m diameter. The cargo dimensions of other launch vehicles are known with less precision. Descriptions of the external dimensions of payload fairings are commonly available, but complicated internal structures and related payload logistics mechanisms would reduce the useable area. Indicatively, payload fairing dimensions of 4 to 5 m diameter and 15 – 26 m length have been advertised.

I provided indicative data on the mass of nuclear reactors, above. Comprehensive data on the mass of most fuel cells was not found and so whether mass, in addition to their consumable fuel requirements, would limit the potential for high-output fuel cells to be launched into outer space could not be confidently determined. However, I located details of the US Space Shuttle's fuel cells. Their total mass was 810 kg plus, for a *three*-day mission, 480 kg each of hydrogen and oxygen (stored as cryogenic liquids) and 600 litres of (generated) water for a peak power capacity of merely 36 kW.[27] I also found an illustration of one commercial fuel cell, a polymer electrolyte membrane cell (PEM) developed to produce 250 kW of electrical power. It displayed a container resembling a large caravan, or 'trailer' in US terminology.[28] This, together with the challenge of launching fuel sufficient to support power generation at levels required by potent space weapons, implies major technological advances would be required before fuel cells could be used in that role.

The development of thin and flexible solar panels suggests neither mass nor physical dimensions would be an insurmountable impediment to launching these panels for subsequent assembly in outer space. Assembling and maintaining the associated supporting framework would be a new and potentially formidable challenge in the history of space engineering.

More generally, modular assembly in space, as has happened with orbiting space stations, might be an option for heavier payloads or for items with dimensions larger than the cargo bays of contemporary or proposed launch vehicles. Any power source or anti-satellite weapon with mass or physical dimensions in excess of size / mass launch capacity, and technologically unsuited to on-orbit assembly, will be constrained to terrestrial operation with projection occurring through the dense atmosphere and into outer space.

Some challenges associated with sustaining weapons in outer space might also be alleviated by on-orbit refuelling or other forms of supplementation. The European Space Agency proposed a vehicle, ConeXpress Orbital Life Extension Vehicle to attach itself to some categories of satellites and provide manoeuvring capability when satellites' fuel supplies were exhausted. An alternative US technology has been demonstrated. Two satellites launched in March 2007 were separated in orbit and then manoeuvred into docking position. Propellant was transferred and components exchanged between the satellites.[29] A variety of supplies has been successfully delivered and transferred to the *International Space Station* (ISS) by US Space Shuttles, Japan's HTV-2, Russia's Progress, and Europe's ATV-2 vehicles. The PRC is developing on-orbit refuelling technology.[30]

A 2009 US report, *Seeking a Human Spaceflight Program Worthy of a Great Nation* (the Augustine Report), concluded that in-space propellant transfer appeared feasible and endorsed additional research, including into the transfer of cryogenic fluids.[31] The Canadian company MacDonald, Dettwiler and Associates, announced in March 2011 its intention to build a spacecraft servicing vehicle. At the same time NASA was conducting experiments into the development of refuelling technologies.

It is nevertheless likely that transporting some devices into outer space will remain challenging for decades despite the potential for increased launch vehicle payloads, reductions in the physical dimensions of some technologies, evolving capacity for on-orbit assembly, and the possibility of on-orbit refuelling.

Notwithstanding all of the challenges of determining, sourcing and launching adequate energy into outer space, some viable space weapon formats have been both developed and demonstrated. Nuclear and kinetic weapons, both of which carry integrated destructive capacity from Earth when they are launched, are now considered. They have already damaged or destroyed spacecraft in orbit. Their availability has global strategic implications.

Chapter 3. Potent Outer Space Weapons that Work

Nuclear Weapons

It is almost certain an isotropic nuclear weapon detonated in outer space will destroy its intended target. During the 1960s, development of nuclear weapons for use in outer space appears to have been motivated by concerns that guidance and control systems then available could not reliably position non-nuclear weapons close enough to their targets for attacks to be effective. [1] Another incentive might have been a perceived need to defend against an incoming swarm of missiles in which circumstance a non-discriminatory destructive response could have appeared desirable.

But the detonation of a nuclear weapon in outer space is also highly likely to generate phenomena that would harm spacecraft in addition to a designated target, and there are now many more such spacecraft than in the 1960s. Some solid fragments from the blast effect could, and probably would, enter orbit as retrograde debris, moving against the direction in which space vehicles in similar orbits normally travel. The additional and enduring challenge of NDRE is described in Appendix 2, *Nuclear-Detonation-Induced Radiation Effects*.

The following review of nuclear weapons primarily involves an investigation into which States can access the weapons as well as launch them into outer space. Only two States, the US and USSR, are known to have launched and detonated nuclear weapons in outer space; in all other cases some element of doubt must exist as to whether States have these capabilities, although the degree of uncertainty varies markedly between States.

I did not locate data on the mass of nuclear warheads attached to missiles. I therefore accepted the mass of nuclear bombs as an indicative substitute when considering the capacity of launch vehicles (aka missiles) to transport nuclear weapons into outer space.

Taking the evolution of US nuclear bombs as a guide, its Mk-1 bomb, yield 15 – 16 Kt, in service until 1950, had a mass of 4 037 kg. The subsequent US Mk-7 bomb with a mass of 747 – 771 kg was reconfigurable to produce a range of yields from 8 – 61 Kt. The US Mk-28 fusion weapon, 70 Kt – 1.1 Mt, in service until 1991, had a mass of 771 - 1052 kg depending on mission-specific configurations.[2]

Other States developing their own nuclear bombs will likely also have started with heavier weapons and incrementally refined them to reduce mass. The unknown and variable factor is whether they subsequently achieved sufficient mass reduction to permit launching their nuclear weapons into outer space, using launch vehicles or missiles available to them, or are likely to do so before about 2030 CE.

I have assumed a State possessing both a long-range ballistic missile and a compatible nuclear weapon will also have developed associated guidance and control systems sufficient to position the weapon so its detonation would result in destruction or irrecoverable damage to one or more target spacecraft.

As a guide to launch capacity, missiles or other launch vehicles are commonly able to launch *vertically* to approximately half the distance they can travel between the place of launch and the place of terrestrial impact.[3] For launches intended to place an object into near-Earth orbit, as contrasted with merely reaching a maximum altitude, the achievable altitude would be *less* than half of the ground-to-ground range because of the requirement to launch at an angle thus allowing the launched object to acquire momentum needed to maintain a stable orbit.

The US and USSR/Russia understand the potential applications for and limitations of nuclear weapons in outer space, both States having built and used them there. In 1958 the US conducted its first nuclear detonations known to have been in the outer space region, as defined in this work. Three tests were detonated at altitudes of approximately 161, 293 and 750 km respectively. The launch vehicles were probably released from USS *Norden Sound, or Norton Sound,* with the test series named Argus Tests. Among the findings from those tests was that the effect of NDRE caused by nuclear weapons can be persistent.[4]

Comprehension of NDRE implications did not stop the testing of nuclear weapons in outer space although a number of tests were designed to simulate rather than to detonate nuclear weapons. In October 1959 the United States Air Force (USAF) successfully launched a missile interceptor, possibly from an aircraft, with the interceptor intended to approach within some six kilometres of a satellite. The near interception of satellite *Explorer IV,* or possibly *Explorer 6,* was considered to indicate a successful trial.

On 9 July 1962 the US in a test known as Starfish Prime detonated a 1.4 Mt bomb at an altitude of 400 km. No satellites were targeted and none have been publically recorded as having been destroyed by the blast effect from that test, but the resulting radiation may have damaged up to six spacecraft. One damaged satellite, Britain–US's *Arial 1,* was 7 400 km distant at the time of detonation.[5]

By 1963 the US Defense budget had authorised research into a defensive missile known as Nike-Zeus which carried a nuclear weapon. It was succeeded by an evolved version known as the Spartan. Spartan had been designed to detonate a nuclear weapon above the atmosphere where it would generate energy with an intensity and geographic expanse sufficient to destroy several incoming missiles almost simultaneously.

The consequences of using Spartan, or any related missiles, might have been spectacular. In addition to the effects generated by a Spartan detonation, any incoming nuclear-weapon-carrying missiles could have been equipped to detonate if attacked thus amplifying the blast and other effects from an aggressive encounter. The technology employed to achieve this form of detonation when attacked was then, and is still, available; it is commonly referred to as *salvage fusing*.

Comparatively little is known of the earliest nuclear tests in outer space by the USSR. Its trials of nuclear devices in the higher altitudes might not have begun until 1961. On 27 October 1961 the USSR detonated two nuclear devices at altitudes of 150 and 300 km respectively.

The USSR also developed a dedicated satellite interceptor launched on the SS-9 Scarp missile. It was tested successfully on 19 October 1968, 1 November 1968 and 23 October 1970.[6] Testing was resumed in 1976. It was not clear from sources I consulted whether this anti-satellite weapon was intended to be armed exclusively with a nuclear device or could also have accommodated a conventional weapon.

In the context that long-range ballistic missiles can be detonated in outer space to destroy spacecraft including other missiles, by 1985 the USSR's arsenal included a total of 10 missile types with a total holding stock in excess of 1 398 missiles, most ostensibly capable of operating in space. The comparable US strategic arsenal then included some 1 000 ICBMs and 600 submarine-launched ballistic missiles (SLBMs).[7] Some controversy over precise numbers might arise but it is clear both States held large numbers of these weapons.

Developments involving nuclear weapons able to be deployed to outer space have continued. On 25 May 2011 a submerged Russian submarine was reported as having fired a missile capable of carrying a strategic nuclear weapon in the third of three successful tests for 2011.[8] Russia has also been reported as developing a new 15-warhead ICBM equipped with stealth features to make detection and engagement more difficult.[9]

India detonated its first nuclear weapon during 1974. On 22 May 1989 India launched the Agni I missile, which had a payload capacity of 1 000 kg and a range of 1 400 km, indicating India might have been able to launch a nuclear weapon to intersect the lower levels of Earth orbits. On 15 October 1994 India successfully launched a Polar Satellite Launch Vehicle (PSLV) with a 1 600 kg payload thus demonstrating an expansion of this potential capacity.

India has since tested an Agni 2 vehicle with a maximum range of 2 500 - 3 000 km.[10] That was followed by other long-range missile developments including two successful tests of an Agni 3 missile, range 3 000 – 3 500 km, in April 2007 and May 2008. Those developments were in addition to the construction of a multiple-independently-targeted re-entry vehicle (MIRV), purportedly able to carry several nuclear weapons, as well as designing a '*hit to kill*' missile.[11] India has also been developing a Geosynchronous Launch Vehicle.

On 25 November 2013 India sent to Mars a 1 337 kg Mars Orbiter Satellite using its PSLV. Given the payload launched and the intricate flight path required to travel to Mars, this event left minimal doubt India could launch a nuclear or other weapon, within the size and mass limitations of its launch vehicles, and acquire a target travelling in ballistic orbit.

Pakistan tested its first nuclear weapons on 28 May 1998. The yields were estimated from seismic analysis to have been approximately 5 to 10 Kt. Since the early 1990s Pakistan has possessed the M-11 (or CSS-7/DF-11) a missile with an 800 kg payload. It subsequently began developing additional missiles, or launch vehicles, but with limited success. Pakistan later gained access to more powerful missiles, including the PRC-originated Shaheen II with a claimed range of 2 500 km and a payload of 1 000 kg. The Shaheen II closely resembles the Chinese missile PRC-M-18. However there is a significant discrepancy between the reported performance of Pakistan's Shaheen II and the M-18, implying the performance of Shaheen II might have been overstated.[12]

Pakistan also manages a missile program based on variants of the DPRK's No Dong technology, sometimes referred to as '*Nodong*'. The missiles have been referred to in Pakistan as Ghauri I, II and III with alleged ranges from 1 500 – 3 000 km. Claims have also been made that one of the Ghauri missiles is '*nuclear capable*' although reports of missile payload capacity found were imprecise and appeared unreliable.[13] Taking account of discrepancies in sources found and the difficulty encountered in verifying data, it was not clear that Pakistan had reduced the mass of a nuclear weapon to a level where it could be carried by any missile to which it had access.

It was also unclear if Pakistan had or could gain permission to operate its more capable missiles independently of the DPRK or the PRC, particularly in an anti-satellite capacity where the aftermath of a nuclear detonation in space would have global implications. Pakistan cannot be disregarded as a future source of space-capable nuclear weapons and associated launch systems, but the evidence is unclear as to whether they exist in Pakistan now or are likely to be developed within the next 20 years.

France detonated its first nuclear weapon in 1960. By 1992 that first test had been followed by 191 more, none of which are known to have been in outer space, although the level of activity indicates France acquired detailed comprehension of nuclear weapons technology. France has also built long-range ballistic missiles able to deliver nuclear weapons. Its submarine-launched M45 missile has a claimed range of 4 000 km and can carry six nuclear warheads.[14] An enhanced successor missile, the M51, was tested in 2010 although performance specifications were not located. France additionally hosts the physical headquarters for and is the majority shareholder in Arianespace, a European company which builds space launch vehicles. The most powerful of Arianespace's launch vehicles is the Ariane 5 with a payload-to-low-Earth-orbit capacity of 20 tonnes.

There is minimal doubt France either has or could rapidly acquire the capacity to launch a nuclear weapon into outer space and to detonate it there.

The PRC exploded its first nuclear weapon on 16 October 1964, launched its first missile designed to carry a nuclear-weapon on 25 October 1966, and detonated its first fusion bomb (terrestrially) on 14 June 1967. The PRC launched its first satellite in 1970 and has possessed long-range missiles capable of lifting a nuclear weapon into outer space since the mid 1980s, or possibly earlier, when the Dongfeng 5 missile was developed. Dongfeng 5 had a range of approximately 12 000 km and a payload capacity of 3 000 kg. In 2013 the PRC probably had nine missiles able to deliver a nuclear weapon into a low-Earth orbit or beyond and maintained ~ 155 nuclear warheads.[15]

Although it has not apparently been demonstrated, the PRC almost certainly has capacity to launch nuclear weapons into outer space and to detonate them there.

Britain first detonated a nuclear weapon on 3 October 1952. But the only British vehicle to have delivered a satellite into orbit is the Black Arrow which launched a 70 kg satellite, once, on 28 October 1971. Through its nuclear-weapon-armed Trident missiles, purchased from and serviced by the US, Britain may have acquired a space-capable weapon. The missiles were carried by British submarines, and so their 7 400 – 12 000 km range was both extensive and flexible. The ballistic trajectories associated with those ranges would require the missiles to travel through outer space. In 2010 Britain's declared nuclear warhead arsenal totalled 225 warheads.[16]

Ostensibly, Britain has the capacity to launch a nuclear weapon into outer space and to detonate it there. From sources available, it was not clear whether Britain has ever had, or could gain from the US, a capacity to act unilaterally in this regard.

Evidence that Israel has nuclear weapons is overwhelming although details of the date of first detonation remain unavailable and no reliable data on Israel's nuclear weapon stockholding was located. One unconfirmed 1999 estimate was that Israel held approximately 400 nuclear weapons but numerous different and also unconfirmed estimates are available.[17]

A report based on information from 1996 included data on Israel's Jericho II vehicle, sometimes referred to as the Shavit. It was then categorised as a space-launch technology. Jericho II's claimed range was 1500 – 4000 km, indicating either some doubt regarding the vehicle's propulsion capabilities or pointing to a wide range of potential payloads. A separate report recording the test of a Jericho referred to it as a ballistic missile.[18] A weapons monitoring organisation, the Wisconsin Project, has estimated that the Jericho II missile could transport a payload of 750 – 1000 kg with a range '*considerably more than 1 500 km*'. A later version, the Jericho III missile, has been widely reported as under development and may have entered service in the period 2008 – 10.

One of many unknown factors affecting Israel's capacity to launch a nuclear weapon into outer space is whether Israel has successfully reduced the mass of its weapons to match the lift capacity of its available missiles. In this context Israel's capacity to conduct tests has been constrained by its policy of opaqueness regarding nuclear weapons. However, its relationship with the US might have facilitated the transfer of relevant data and related technologies, or actual weapons, to an extent invisible to unclassified sources.

Israel's capacity to utilise nuclear-armed space weapons remains uncertain, although cumulative evidence suggests it might be able to launch a nuclear weapon into low earth orbit and detonate it there. If it cannot do so in 2014, Israel's focus on continually developing its military capabilities suggests it would have this capacity within one to two decades.

Available information and data on Iran's launch capacity is complicated and confusing. Iran might have developed the Shahab-3 missile or space-launch vehicle which has a reputed payload of 1 200 kg and a range of 1 300 km. The device resembles a DPRK No Dong 1 missile. There might also exist an evolved version, the Shahab-4. Subsequent versions referred to as Shahab-5 and 6 have also been reported but information and data regarding their alleged performance appeared in numerous versions and was not reliable.

Iran's missile inventory also includes the Sejil missile also known as the Ashura. Another and almost indistinguishable vehicle, referred to as the Safir, has been used by Iran to launch small satellites. Whether the Safir/Sejil/Ashura could be modified to launch a nuclear weapon into outer space has not been confirmed. One report claimed the Safir-IB had a vertical range of only 300 – 450 km with a 50 kg payload; another claimed that the range was 2 000 km, payload not specified. Other possible launch vehicles in Iran's expanding inventory include the Kavoshgar, vertical limit 120 km, payload not identified, and possibly the Simorgh, vertical range 500 km and payload 100 kg, and the Ghadr. The name Simorgh has also been associated with the Shahab 5 missile.

During 2013 publicity surrounded Iran's launch into space of monkeys on two occasions. The absence of technical detail made evaluation difficult but it appeared the launches were almost vertical, to 120 km, and would therefore have achieved a lower altitude had they been angled to permit entry into orbit. Although the analogy is chronologically remote, the USSR's *Sputnik II* which carried one dog into space had a mass of 508.3 kg. If the mass of Iran's monkey capsule was comparable, then significant progress in launching spacecraft may have been achieved by indigenous developments.

To launch a nuclear weapon into outer space, Iran would have to build or acquire a weapon having a mass compatible with the launch capability of one of its launch vehicles. It is not possible to predict the likelihood of Iran acquiring any such weapon because of unknown factors involving Iran's nuclear-technical capacity. On 24 November 2013 an agreement was reached between Iran and a consortium involving Britain, the PRC, France, Russia, the US and Germany to restrict Iranian nuclear enrichment and to permit inspections of its nuclear facilities. By early 2015 it was not clear if the agreement would be implemented to the satisfaction of all involved parties.

During the mid 1990s the DPRK acquired a Russian launch vehicle designed in the 1960s, the R-27, which it modified and renamed Musadan.[19] Although technologically old, the Musadan vehicle had a ~3 200 km range and could carry a 1 200 kg payload. Performance at that level would have been adequate to launch a technologically refined nuclear weapon into outer space, if the DPRK had possessed such a weapon. (The DPRK first tested a presumably high-mass nuclear weapon on 9 October 2006.)

The DPRK then developed and modified several missiles. However, claims regarding the capacities of its missile fleet vary significantly. As at November 2012, the DPRK's most recent three missile tests had all failed, marking that State as the only nuclear-armed nation with a missile program that appeared to have *reduced* its capacity to deliver warheads. However on 12 December 2012 the DPRK used its Unha rocket to launch a satellite into orbit. Details of the satellite mass or other specifications were not located, although the claim it was broadcasting a patriotic song implied a small and relatively uncomplicated vehicle.

It was not possible from data available to determine whether the DPRK might be able to launch a nuclear weapon into outer space. Minimal information was found concerning either the mass or other specifications of the DPRK's nuclear weapons, or to support any assessment of any recent launch vehicle enhancements. Whether the successful launch on 12 December 2012 was serendipitous or represented a sustainable enhancement of capability remained unclear.

It is apparent the DPRK is continuing to progress its nuclear weapons program, the State having conducted a third nuclear test on 12 February 2013 regardless of international pressure to not do so. Despite a depressed economy, the DPRK has continued to invest heavily in military capabilities. Consequently, it might become able to launch a nuclear weapon into outer space within the next 20 years.

Japan has refused to acquire nuclear weapons, although continuity of this policy cannot be guaranteed, despite Japan's history of formally supporting nuclear non-proliferation initiatives.[20] The Japanese Constitution does not prohibit the State from acquiring or using nuclear weapons for self defence but rather renounces war as a sovereign right of the nation as well as renouncing the threat or use of force as a mean of settling international disputes. These constraints have not prevented Japan's formation of a self-defence force, although the additional move to acquire a nuclear armoury would be a radical departure from Japan's consistent post Second World War policy. Japan's nuclear reactors nevertheless provide a foundation from which the raw material for nuclear weapons could be concentrated.

Japan launched its first satellite in 1970 using the Lambda L-4S vehicle. Subsequent launch vehicles have included the M-V, capable of launching a 1 800 kg payload into low Earth orbit, and the Epsilon, with a 1 200 kg payload capacity. The launch vehicle built by Japan to resupply the ISS, the Kounotori/ HTV-3, has a payload capacity of at least 3.5 tonnes. Any of these launch vehicles could transport a modern nuclear weapon into outer space and into orbit.

Given Japan's apparent aversion to possessing nuclear weapons, it is unlikely a capacity to launch them into outer space will be developed unless there is a radical change to announced Japanese policy. Conversely, Japan could be incrementally reinterpreting its constitutional constraints. Japanese vessels have been deployed to waters adjacent to the Senkaku / Diaoyu Islands, a region involving a territorial dispute with the PRC. Prime Minister Abe has been reported as intending to revise Japan's pacifist stance.[21]

Brazil began investigating nuclear fission in the 1930s, later embarking upon a nuclear weapons program. The program was discontinued in 1990 but Brazil's technological base, its experience with managing nuclear reactors, and its retention of the Resende II uranium enrichment facility all point to a capacity to build nuclear weapons.

The capacity to launch a nuclear weapon into outer space using an indigenously developed missile is not now resident in Brazil. Its most ambitious launch vehicle, the Veículo Lançador de Satélites, has experienced several launch failures. Brazil's most recent (2014) attempt to acquire a launch capacity other than for micro satellites has been an arrangement with Ukraine to launch a Cyclone-4 vehicle. If it performed to design specifications the Cyclone-4 would carry >1 000 kg to an orbit of ~ 800 km. As at mid 2014 a major rift between Russia and Ukraine cast doubt over whether those arrangements would be able to proceed as originally planned.[22]

The combination of past failures and reprioritisations suggests Brazil is unlikely to acquire a significant indigenous launch capacity by 2030. It might acquire a substantial launch vehicle from Ukraine or elsewhere but this is not assured. No evidence located suggested Brazil was again attempting to develop or acquire nuclear weapons or had any ambition to use them as weapons in outer space.

In summary, precedents attributed to the US and USSR-Russia demonstrated nuclear weapons can function effectively in outer space to damage or destroy orbiting spacecraft. Additionally, the PRC, France, India, and Britain almost certainly have the technological capacity to launch and detonate nuclear weapons in outer space, although no record of them attempting to do so was found. Pakistan, Israel, Iran, Brazil, Japan and the DPRK could acquire this capability within the next two decades, although the likelihood of any one of them doing so will vary significantly between States.

Detonation of a nuclear weapon in outer space might under some circumstances be regarded by a State as justifiable, particularly in a defensive context. The weapon would not kill or mutilate large numbers of people, as could occur with any such terrestrial event. Resultant property damage in outer space would be almost insignificant compared against the financial costs and related losses occasioned by destruction of a terrestrial city. The damage to terrestrial infrastructure from space-sourced NDRE could be, in both financial and technological terms, minor compared against the harm inflicted by a nuclear weapon detonated in the dense atmosphere above a populated city.

Kinetic Weapons

The concept of a kinetic weapon suitable for use in outer space is applied loosely in this work. Some weapons thus categorised have not inflicted damage solely through direct physical impact but have carried explosive charges or other destructive capabilities intended to be activated in the immediate vicinity of their targets. However, outer-space weapons described as *kinetic* have invariably been close-quarters weapons, weapons capable of inflicting physical damage other than through electronic techniques or using fissionable materials. I have adopted that broad approach to kinetic weapons.

The USSR developed at least one anti-satellite kinetic weapon during the 1960s. The Polyot satellite interceptor employed kinetic energy to destroy satellites in low-Earth orbit.[23] The primary elements of this system had been built by 1967 and its first satellite interception mission appears to have been accomplished in October or November 1968. The weapon was allegedly capable of destroying spacecraft at altitudes from 250 to 1000 km. It was accepted for military use with the designation IS-MU and remained in service until 1993.

In 1974 the USSR tested a Nudleman rapid-fire cannon carried on the *Salyut-3* space station. During the 1980s the USSR also sponsored anti-missile or anti-satellite programs designated SK-1000, D-20 and SP-2000. These programs were '*formalised*', possibly meaning accepted for further development, in 1985. The USSR also considered using its R-36M, UR-100H, SS-18 and SS-19 intercontinental ballistic missiles for deployment against space vehicles, although these potential encounters might have been nuclear.

A US Congressional Staff Report from 1961 refers to a 1960s US anti-satellite weapon, the Early Spring, designed to hover then intercept a designated satellite and destroy it by use of a '*mechanism*'. That weapon, which might have been a fragmenting space-mine, had been developed by the US Navy.[24] The US also developed a '*fractional orbit bombardment system*', although it was not clear from references consulted whether the weapon described was intended for space-to-space or space-to-Earth engagements or both. The US began developing an additional device referred to as an anti-satellite miniature vehicle.[25]

In 1977 the US began testing its Miniature Homing Vehicle. The vehicle could be fired from an aircraft, designated F-15. It used an SRAM-Altair booster which had a claimed maximum altitude of 1 000 km. On 10 June 1984 the US employed the Miniature Homing Vehicle against a Minuteman missile warhead, which it destroyed, thus implying both an anti-missile and an anti-satellite capability.

On 13 September 1985 the USAF again used a kinetic weapon to intercept its own satellite, the *P78-1 Solwind,* at an altitude of 515 km or possibly 600 km. The test created 285 pieces of detectable debris. Additional tests were conducted in August and September 1986.

The most ambitious anti-missile or anti-satellite weapon ever proposed for launch into outer space was a component of the US Brilliant Pebbles program, forming part of President Reagan's 1983 Strategic Defense Initiative. Brilliant Pebbles was to include between 1 000 and 4 600 satellites (estimates vary between sources), each hosting a kinetic space weapon based on a satellite known as the Kinetic Energy Antisatellite (KEAsat). Despite an apparently promising beginning, funding for the program was cut almost to the point of elimination in 1993.

But enthusiasm for Brilliant Pebbles might not have evaporated along with its funding. The KEA-sat could have been resurrected in 1996 and subsequently funded until at least 2005 at which time it formed part of the Applied Counterspace Technology testbed located at Redstone Arsenal.[26] In 2004 a subsequent kinetic weapon program, the US Near-Field Infrared Experiment (NFIRE), was claimed to resemble aspects of Brilliant Pebbles.

Additionally, the US Missile Defense Agency in 1993 proposed a Space-Based Boost program intended to comprise a constellation of orbiting '*kinetic-kill*' vehicles designed to eliminate intercontinental ballistic missiles in their boost phase. However, this might have been primarily in the space-to-Earth weapons category and so beyond the defined limits of this manuscript.

Until 2007 the US allegedly conducted other kinetic interception tests in outer space using a ground-based system, Terminal High Altitude Area Defense (THAAD), although available literature commonly asserts its vertical range to be ~ 150 km. A sea-based anti-ballistic missile system, the Aegis Combat System has also been linked to satellite interception missiles.

In the first decade of the 21st century the US continued to develop kinetic weapons with anti-satellite applications. Most such programs were pursued under the rubric '*ballistic missile defence*' and technologies developed were intended primarily to achieve that outcome although some capabilities would be transferrable to engage more predictable targets such as satellites.

A US ballistic missile interception test conducted on 1 September 2006 was successful in that the designated target was obliterated. On 20 February 2008 the US Navy demonstrated an anti-satellite capability when it intentionally destroyed a satellite, *US-193,* at an altitude of approximately 240 km using a Standard Missile Type 3 (SM-3). The SM-3 is commonly described as an anti-missile weapon.

Before the 2008 interception event General James Cartwright, Vice Chairman of US Joint Chiefs of Staff, was reported as stating that the SM-3 missile had been modified in a manner '*not transferable to fleet configuration*', although no explanation appears to have been offered in support of this caveat.

One unnamed Director of the US Missile Defense Agency is claimed to have stated that the Agency planned in 2008 to begin experiments in space to test the viability of placing anti-missile interceptors in orbit. One such program, the Space-Based Interceptor Test Bed, would launch up to five devices capable of shooting down missiles. Another reference to the same or a similar program appeared during 2005 when a constellation of space-based missile interceptors was mentioned.[27]

The US was also reported to have tested a Demonstration of Autonomous Rendezvous Technology (DART) in April 2005 resulting in the test vehicle colliding with its target satellite. This might have been a test of an Experimental Spacecraft System (XSS) -11 reported by the *New York Times* and apparently intended to demonstrate a satellite interception capability.[28]

On 17 August 2010 the US Raytheon company claimed it had, on 6 June 2010, demonstrated successfully a new '*exoatmospheric kill vehicle*'. The weapon was described as having been designed for '*midcourse*' application against ballistic missiles using the impact force of hit-to-kill. In other words, it was a space weapon designed to apply kinetic force and one equally or better able to engage satellites.[29]

In December 2011 the space media announced a contract had been awarded to Raytheon by the US Missile Defense Agency for continued engineering design and development work on the SM-3 Block IIA missile, described as a co-development effort between the US and Japan. The evolved missile, with larger rocket motors than previous models and an advanced kinetic warhead, was claimed to be on schedule for deployment in 2018.

Russia, the US and conceivably other spacefaring States have expanded the concept of kinetic intervention by developing a rocket able to attach itself to orbiting spacecraft. It could then be operated to slow the targeted spacecraft until the target was forced to re-enter the dense atmosphere, being destroyed in the process.

On 23 March 2001 Russia employed a well-publicised example of this technique by using its Progress M1-5 rocket to de-orbit the *Mir* space station. This event confirmed Russia's capacity to intercept and de-orbit the largest of space vehicles operating in low Earth orbit. However the Progress required a total engine burn-time of 44 minutes to achieve success, pointing to some limitations in its application as a weapon. Further development of the capability appeared to be technically feasible.

With minimal publicity the US had already demonstrated a similar technique in 1992 when crew from the space shuttle mission STS-49 (7 – 16 May 1992) attached a rocket motor to a malfunctioning satellite (*INTELSAT VI*) allowing it to be boosted into a higher orbit. A similar intervention could have been used to deorbit a similar satellite.

The PRC demonstrated a kinetic anti-satellite weapon in January 2007. On 11 or 12 January 2007, the 954 kg *Fengyun-1-C* satellite was destroyed by a PRC-operated missile at an altitude of ~864 km over central China. I did not find any record of the impact being observed but a Dong-Feng 21 missile was launched from the Xichang Satellite Launch Centre at the precise time required to achieve interception with that satellite. PRC airspace related to the missile launch had been reserved in advance. At the time an interception would have taken place the *Fengyun 1C* disappeared and was replaced by a cloud of detectable debris.

On 11 January 2010 the PRC launched a missile from its Korla Missile Test Complex and intercepted a CSS-X-11 ballistic missile launched from Shuangchengzi Space and Missile Centre. The interceptor was either designated SC-19 or was a component of a broader anti-satellite system designated SC-19. This additional demonstration confirmed the PRC's capacity to eliminate satellites in low Earth orbit and to counter some components of a missile attack from outer space. However, the lack of available data, including target velocity relative to the interceptor and angle of interception precluded an assessment of how potent the PRC's capability had become.

Referring to India, one analyst recorded:

> *'... in November [2006] India used a kinetic interceptor mounted on a Privthi missile to intercept another Privthi missile at 50 km. India's Defence Research and Development announced the test as the first step in developing an exoatmospheric interceptor.'* [30]

This claim was congruent with other developments in Indian counter-space technology and pointed to India advancing toward acquiring an anti-satellite kinetic weapon within the present decade. On 6 March 2011 India reported the successful test of an air defence interceptor equipped with radio frequency-seeking guidance as well as directional warheads.

According to the Head of India's Defence Research and Development Organisation, V Saraswat, the interceptor was a component of technologies which could be used for anti-satellite missions.[31]

It is difficult to ascertain whether Pakistan has developed or will develop anti-satellite kinetic weapons in the foreseeable future. Pakistan's capacity to master the relatively less complicated technologies associated with launching a nuclear weapon into the vicinity of an orbiting spacecraft is in doubt. No evidence was found of Pakistan attempting to develop an indigenous anti-satellite kinetic weapon although it might, with support from the PRC, import that capability. Under those circumstances, further and unanswerable question must arise as to whether the PRC would permit its technology to be used independently of its own strategic policy.

Japan has been moving incrementally toward acquiring kinetic weapons deployable against satellites and missiles. On 19 December 2003 the Japanese Cabinet agreed to a policy proposal, *On Introduction of Ballistic Missile Defense System and Other Measures*. Japan's anti-missile armoury includes the SM-3 missile. The potential anti-satellite role of that weapon when in Japanese hands was demonstrated in December 2007 when an SM-3 launched from the Japanese ship *Kongo* destroyed a missile resembling a DPRK No Dong. The interception occurred at an altitude of approximately 160 km. Relevant cooperation between Japan and the US is continuing.

The DPRK has been more secretive even than the PRC, thus a reliable determination of its capacity to deploy a kinetic anti-satellite weapon could not be made. The DPRK's missile-related capacity, as demonstrated by attempts to launch satellites, implies it does not (2015) have the technological capacity to build a reliable kinetic anti-spacecraft weapon and navigate it with the necessary precision toward a target in ballistic orbit. Given its focus on increasing its military capacity the DPRK might be able to do so within 15 years.

Israel is unlikely to encounter major technical challenges in building upon its existing missile defence program to produce a kinetic weapon for use in outer space. The US and Israel have been reported as cooperating on installing an '*upper tier*' missile defence architecture including using the Arrow 3 and SM-3 missiles. Israel's Iron Dome missile defence system is being supported by the US. Israel is experienced in building UAVs including the Hermes 450, a multi-mission aircraft with flight endurance of '*over 20 hours*', illustrating mastery of precision guidance and remote control technologies.[32]

Iran's achievements in missile technology have not been welcomed globally. The Shahab-3 missile appears to be based on the DPRK's No Dong although Iran is claimed to be experimenting with enhancements to the missile's range and accuracy. Given past technological achievements Iran could conceivably build an effective anti-satellite kinetic weapon within 15 years. Whether or why Iran would be motivated to do so is not known.

There is close cooperation between European members of NATO and the US in matters associated with missile defence. By association and implication these arrangements could extend to anti-satellite capabilities for kinetic weapons. Under the US '*phased adaptive approach*' missiles installed in European facilities will include the SM-3, variants of which have been used to intercept and destroy satellites.[33]

Evidence indicating any Eastern or Western European State is pursuing an independent and indigenous program to develop an anti-satellite kinetic weapon was not located. Russia has been considered separately. Given its technological achievements with space launch and missiles more generally, France appears to have greater capacity to do so than any other European State.

When the PRC used a kinetic weapon to destroy its own *Fengyun-1-C* satellite at an altitude of ~864 km this became the highest altitude at which a publically known attack by a kinetic weapon had occurred. However, the precision with which some spacecraft have been able to approach, land upon or damage asteroids and comets indicates capacities exist to attack any satellite in any Earth-centric orbit.

During July 1999 the US spacecraft *Deep Space 1*, passed asteroid Braille at a distance of 26 km. In December 2000 the US spacecraft *NEAR Shoemaker* landed on asteroid 433 Eros. In July 2005 *Deep Impact*, another US spacecraft, fired a '*smart impactor*' into the path of comet Tempel 1 creating a crater later estimated as 100 m wide and 30 m deep. Japan's *Hayabusa* intercepted the Itokawa asteroid, landed on it twice and attempted to collect a geological sample during November 2005. In December 2012 the PRC's spacecraft *Chang'e-2* passed the asteroid Toutatis at a reported distance of 3.2 km. During November 2014 the European Space Agency's *Rosetta* landed on the comet 67P/Churyumov-Gerasimenko. In 2014 Japan was testing a '*Small Carry-on Impactor*' designed to be transported to an asteroid where it would create an '*artificial crater*'.

In summary, kinetic devices began to demonstrate their effectiveness as weapons in outer space following the Aerobee concept demonstration by the US on 16 October 1957. They were developed and have been demonstrated by the US, Russia, PRC, Japan (with support from the US) and possibly by India. The suite of technologies needed to use such weapons against satellites may have proliferated to additional States. Acquiring the capacity to deploy these weapons against satellites could be less challenging than using them to attack ballistic missiles, given the predictably regular orbits of satellites.

Several States have developed the capacity to kinetically attack spacecraft in any Earth-centric orbit as illustrated by scientific missions involving asteroids and comets. A capacity to attach rockets to orbiting spacecraft so the spacecraft are forced to re-enter the dense atmosphere has also been developed and demonstrated. Kinetic anti-satellite weapons can be launched terrestrially or from vehicles already in outer space.

Space-Mines

Literature examining weapons in outer space has described technologies designed to release a cloud of objects intended to collide with, or be struck by, orbiting spacecraft. The devices, commonly termed 'space-mines', would cause damage through kinetic energy. They can be divided into two broad categories: mines designed to place a cloud of objects into orbit and mines intended to launch objects to temporarily intersect the path of an approaching spacecraft. Either category could cause devastating damage to spacecraft, but consideration of space-mines has been generally confined to the 1950s or early 1960s when the density of US and USSR-launched satellites could be counted in the tens as compared against 2010 and beyond with more than 1 000 active payloads in orbit and ownership extending to more than 50 nation States. [34]

From a technical perspective, the challenges of launching a cloud of objects to temporarily intersect the path of a specific spacecraft (viz. adopting a trajectory from which the pellets would soon fall back to Earth) are formidable. The concept assumes an attacker with the capacity to deliver pellets to a precise position at the exact time a target spacecraft would be present. The pellets released would need to be sufficiently numerous to ensure one or more collisions with the target and the target could not have been designed to resist the impact thus caused. Taking all of these factors into account it is likely an aggressor having the technology necessary to mount this form of attack could also deploy a small homing missile or other kinetic device with a greater likelihood of success.

Activation of a space-mine with its payload designed to enter into orbit would now have global implications in both the geographic and the political contexts. At an approximate altitude of 900 km or higher the pellets could remain in orbit for centuries where they would encounter orbiting spacecraft unpredictably and without discrimination as to the State of spacecraft ownership. Given all major spacefaring States would be actually or potentially affected by orbiting debris from a space-mine, an emphatic and possibly coordinated terrestrial retaliation against the perpetrator could be anticipated.

The era in which space-mines might have been viable as weapons in outer space has almost certainly passed. Other and more reliable outer space weapons are available. The concept is not considered further.

Chapter 4. High-Power Ray and Beam Technologies

The application of beamed-energy as a potential space weapon has attracted global interest. This chapter reviews the related concepts with the objective of determining whether published optimism has been, or seems likely to become, translated into technological achievement. The energy requirements and related parameters identified in Chapter 2 apply here also.

Laser Weapons

Laser weapons have been militarily attractive because energy would be transmitted at the speed of light (*c*) so that no target at which such a weapon was fired could manoeuvre to avoid an initial encounter. The challenges of achieving such a transmission through the outer space region at a physically destructive level are significant. Some of them I have already referred to, in particular the quantum of power needed to support such devices even if they involved beams of only a few centimetres in diameter. Nevertheless, numerous techniques based on employing lasers as space weapons have been conceptualised and attempts made to apply them.

There is no doubt at low power the beams from lasers can dazzle or permanently damage optical and related sensors carried on satellites and this is considered further in Chapter 5. At higher power, lasers might be able to overheat spacecraft by increasing the internal temperature of target vehicles, or their integrated solar panels, at rates faster than the spacecraft could process, leading to component failure. At very high power, space-based lasers could, at least in theory, melt or oblate satellite sheaths and burn their electronic or other components.[1]

One of the first and still extant challenges associated with using lasers as high-powered space weapons concerns mirror dimensions. Broadly, for most categories of laser the transmission of a powerful and focussed beam requires an integrated mirror. Providing perspective to the dimensions of mirrors, monolithic mirrors of 8.4 m diameter have been designed for the international (terrestrial) Giant Magellan Telescope. A segmented 10.4 m diameter reflector used by the terrestrial *Gran Telescopia Canarias* is among the largest telescope reflectors ever constructed.

For conceptual high-powered lasers, a mirror of 10.0 m diameter has been estimated to produce a beam spot of 0.3 m diameter (covering a plane area of ~707 cm2) at a distance from its target of 1 000 km, provided all other factors were functioning at a theoretical 100 % efficiency.[2] A similar laser in geosynchronous orbit (GEO) and directed at a modest angle toward satellites in the lowest orbits, a nominal intervening distance of ~40 000 km, was estimated to require a perfect mirror of approximately 130.0 m diameter to focus a beam of 1.0 m diameter on its target.[3] The beam would then cover a plane area of ~7 854 cm2 . There are obvious implications for production capacity as well as for launching these components into outer space, even if such mirrors were segmented rather than monolithic.

Additionally, the difference in surface area between a 0.3 m diameter and a 1.0 m diameter beam-on-target is obviously greater than one order of magnitude. Therefore, deductively, delivery of the same concentration of energy over the 40 000 km intervening distance, instead of 1 000 km, would demand more than 10 times the projected energy or require a dwell-time on target more than 10 times as long. This dwell-time assumes a target oblivious to being attacked, not implementing any protective response, and not expelling accumulated energy through radiation. Any defensive response by the target space vehicle would require the laser weapon to deliver yet more energy or to dwell on the target for a still longer period of time, noting also a satellite in its low orbit would be travelling at more than twice the velocity of a GEO-based weapon and thus disappearing from view.

As a generalisation, most of the power provided to a laser is converted into light (broadly categorised) and heat. In outer space the heat would have to be dispersed by radiation to prevent accumulation and eventual component failure. Heat would create a signature detectable in the outer space region thus providing an opportunity for a laser weapon to be identified and categorised as a target.[4]

As with other weapons, every laser weapon located *in* outer space must be enabled by energy launched with it or acquired subsequently in outer space. The primary alternative is to build terrestrially-based lasers and then beam energy through the dense atmosphere toward targets already in outer space. On Earth, or even in laser-equipped aircraft, the acquisition of fuel and therefore the challenge of energy replenishment would be relatively uncomplicated.

Laser technologies allegedly able to function either as anti-satellite weapons in outer space, or to operate as anti-satellite weapons from Earth into outer space, include: chemical, X-ray, excimer, diode-pumped, and free-electron. Other laser formats might have been developed clandestinely for intended space applications as weapons.

Laser developments reviewed here focus on attempts to build high-powered and physically destructive weapons. Several States, but primarily the US, have attempted to develop space weapons based on laser technologies. Discussion following refers to either the constant or mean output levels of lasers considered. However, capacitors and analogous technologies can provide fundamentally higher levels of pulsed output power in some cases. Amplifying technologies include Q-switching and master oscillator power amplifiers (MOPAs). Some lasers intended to project ablation effects are designed to utilise pulsed power.

In relevant literature the laser output is commonly shown as kW, implying a continuous wave (CW) format although literature describing some specific lasers did not identify them as either pulsed or CW. Consequently data following should be accepted as indicative in this regard.

Lasers employing chemical reactions were among the earliest technologies considered for application as beamed space weapons. Compared against other laser categories, chemical lasers were relatively simple and could be scaled to high output levels.

In the 1980s US analysts believed the USSR already possessed ground-based lasers able to interfere with satellites. Developmental activity was believed to be occurring at a test site based at Sary Shagan (in Kazakhstan, sometimes written as 'Saryshagan').[5] Within the US, several organisations were developing high-powered lasers intended for military applications including applications in outer space.[6]

Initial attempts at constructing a high-powered chemical laser were crude by 21st century standards. One terrestrial (US) chemical laser scheduled to be completed in the 1990s was estimated to consume 100 MW of power, require more than 1 703 400 000 litres of water per annum for coolant and absorb support from an operational staff of 250 people.[7] An earlier version was reported as having required 370 people and 34 000 litres of water as coolant to operate for a few seconds.[8]

Despite limitations, developments of chemical laser weapons continued, including construction of the US's (terrestrial) Mid-Infrared Advanced Chemical Laser (MIRACL), described as a laser-firing anti-satellite system.[9] Officially endorsed technical details of this laser were not located although the Federation of American Scientists website claimed MIRACL operated in the band 3.6 -4.2 μm and had a maximum lasing duration of 70 s.

The first allegedly successful MIRACL tests against a satellite were performed in 1997. A subsequent assessment claimed this laser could disable solar panels or damage the optical equipment of satellites orbiting at altitudes of between 400 and 700 km.[10] An evaluation of those claims was impossible without access to additional information. However, being ground-based, MIRACL would likely have encountered the underside of structures supporting solar panels. An analysis of material therefore actually damaged in order to disable the solar panels, was not revealed in reports found.

By 2001, the US was investigating the Evolutionary Aerospace (sometimes presented as 'Air and Space') Global Laser Engagement, or EAGLE, a hypothetical project in which terrestrial lasers as well as space-based lasers would aim at high-altitude relay mirrors. The mirrors would refocus and retransmit the laser energy, directing it to destroy missiles or other targets. EAGLE may have been part of the Integrated Flight Experiment plan designed in part to address several emerging issues related to lasers in space. It envisaged a self-sufficient space-based laser with '*an 8 – 10 meter monolithic or segmented mirror and … a launch weight of some 80 000 pounds* [~ 36 290 kg]'.[11]

The US was almost simultaneously developing a chemical oxygen-iodine laser (COIL). One such laser was mounted in a modified Boeing 747 aircraft, possibly as a precursor to a space-based laser. Its wavelength was 1.315 μm. One observer has claimed this laser could hit ballistic missiles in their boost phase at a range of 200 – 250 km.[12] Another report alleged, on 11 February 2010, the laser heated a '*boosting ballistic missile to critical structural failure*'.[13] The altitude of the encounter, distance from laser to target and target composition details were not found.

The selection of a Boeing 747 platform for the COIL laser may have been revealing in that the payload of a conventional Boeing 747 is approximately 110 tonnes, implying the COIL and its integrated fuel tanks and related equipment had a combined mass of approximately that magnitude, this being despite the availability of periodic terrestrial refuelling between sorties.

Descriptions located of the airborne COIL laser invariably omitted details of the fuel-mass requirement, leading to a suspicion this might have been a major impediment to further development of the technology as a weapon in outer space. A COIL laser requires five fuels: molecular iodine, potassium chloride, oxygen, gaseous chlorine and hydrogen peroxide, although the final two can under some circumstances be regenerated. The implication that fuel mass might have been an Achilles heel has been supported by the Global Security organisation which published a paper claiming fuel for the US airborne COIL required support from two C-17 transport aircraft.[14]

Published assertions regarding COIL laser efficiency are difficult to interpret because of variations in terminology and uncertainties regarding key design aspects of the lasers being described. Estimates of fuel-to-energy output efficiency vary from 23.4% to 40 %.[15] The US Defense Advanced Research Projects Agency conducted a study suggesting *potential* to achieve fuel consumption of 5 kg/kW at the megawatt level.

The US airborne COIL laser program was subsequently cancelled. Taking account of chemical-laser fuel requirements and the physical dimensions of chemical lasers, it is almost certain they will not be employed as high-powered weapons *in* outer space. Deployment of chemical lasers from terrestrial installations remains conceivable but challenges associated with projection through Earth's dense atmosphere will remain and very high levels of energy will need to be projected to achieve more than peripheral damage to spacecraft. The compromise of firing from above most of the terrestrial atmosphere was probably under consideration by the US when developing its aircraft-based COIL, but even this apparently proved impracticable.

In the 1980s it was theorised that X-ray emissions could be focused into destructive laser beams not requiring monolithic or other mirrors. The only viable source of power then identified for an X-ray laser was a detonated nuclear weapon. Any such device could have been used only once.

The US at that time claimed success in generating X-ray lasing during an underground nuclear explosion. Related experiments associated with design concepts were known as Excalibur and Super Excalibur. From sparse available literature it was not clear if progress had since been achieved with the creation of identifiable multiple beams or the directional control of them.

However, production of laser power at X-ray frequency without a nuclear weapon as its power source is very difficult. On 21 April 2009 staff at the Stanford Linear Accelerator Center (SLAC) reported creation of the '*first ever glimpse*' of X-ray laser light produced in a laboratory. The demonstration was described as requiring only one third of SLAC's linear accelerator, the total accelerator length being 3.2 km.[16]

The impression of technology still at the early stage of development was strengthened in January 2012 when the Stanford Center reported using an X-ray laser to heat a '*miniscule piece*' of aluminium for less than one trillionth of one second ($<10^{-12}$s) to greater than 2 x 10^{6}°C.[17] According to results published by Lawrence Livermore National Laboratory, the X-ray laser functioned at maximum intensity with a wavelength optimised to 14.7 nm. The '*tabletop*' demonstrator (physical dimensions not specified) could be fired '*every three or four minutes*'.

Developments necessary to produce a viable X-ray laser for use as a weapon in outer space, other than by using a nuclear weapon, would need to include provision of an adequate power source. Additionally, the technology would have to be within the mass and physical dimensions constraints permitting launch into outer space or subsequent assembly on orbit.

An alternative would be to launch a nuclear-weapon-enabled X-ray laser only when required to combat a swarm of incoming missiles (a 'pop-up' response). Associated challenges would be to prepare for the launch, execute it and enable the carrier missile to travel until within range of attacking missiles, all of those processes having been accomplished after the attacking missiles had been launched and identified as potentially aggressive.

Considering progress not made during past decades and the magnitude of extant challenges it seemed unlikely any State could produce an X-ray laser able to function as a weapon in outer space, before 2030, unless it was enabled by a nuclear weapon.

Excimer, or exciplex, lasers emit pulsed energy. Commonly, the laser output is 10 – 50 ns pulses typically of 0.2 – 1 J. These lasers utilise a range of excimer gain media gases which, in combination with resonator design, determine wavelengths as shown in the table following. Unlike some laser formats the application of which can apply heat at destructive levels, the excimer beam disrupts molecular bonding, resulting in target ablation. This indicates a requirement for preliminary understanding of target composition if the laser energy is to be applied efficiently.

Table 4 – 1. Excimer Gas, Wavelength and Relative Power [18]

Gas	Wavelength (nm)	Relative Power
Argon fluoride	193	60
Krypton fluoride	248	100
Xenon chloride	308	50
Xenon fluoride	351	45
Molecular fluorine	157	10
Krypton chloride	222	25
Molecular nitrogen	337	5
Ionised nitrogen	428	3

Until recently, excimer lasers available for terrestrial applications experienced component corrosion and gas contamination with adverse consequences for component lifetime. Structural integrity was also a challenge with gas pressures commonly in the range 1.5 x 10^5 Pa to 6 x 10^5 Pa. Contemporary manufacturers of terrestrial excimer lasers claim to have resolved these and related technical issues therefore improving longevity and reliability, extending the capacity of excimer lasers to function for periods of between months and years.

Accepting as accurate the manufacturers' claims regarding component reliability and durability, the primary residual challenges to operating a high-powered excimer laser in the outer space region appear to be availability of sufficient enabling power as well as mirror dimensions. Terrestrial excimer lasers have a conversion efficiency, electrical input to optical output, of approximately 2 %. [19]

During 1986 the Los Alamos National Laboratory reported developing a krypton fluoride (KrF) laser intended to be scalable to 100 kJ/s and eventually to 200 kJ/s. Referring to the nominal indicator of 50kJ/cm^2/s as likely to damage many spacecraft, this laser when operating at 200 kJ/s could have supported an effective beam area of 4 cm^2 (radius ~ 1.1 cm).[20]

In contrast with more impressive claims made for lasers, an excimer laser was launched into outer space in March 1988. It provided 0.4 kJ for a duration of 0.5 s indicating that the technology was immature in relation to requirements of a high-powered weapon.[21]

Excimer lasers are now rarely mentioned in literature relating to outer space applications as weapons, although interest in excimer lasers for peaceful purposes remains high. They are most commonly employed for surgery and other precise, short-range, low-powered functions. The possibility that excimer lasers will in the future be employed in the high-powered anti-satellite role cannot be dismissed unconditionally. It is conceivable, but no more, that presently unknown developments might lead to the construction of a space-based excimer laser weapon by 2030.

I have selected the neodymium-doped yttrium-aluminium-garnet (Nd:YAG) laser as representative of the diode-pumped category. This laser commonly functions with a wavelength of 1064 nm, but modifications can enable operation across the range 940–1440 nm. Pulsed and continuous applications are achievable. The input electrical to optical efficiency ratio of Nd:YAG lasers is unclear because a range of evolving influences exists. Following a review of available literature I accepted an indicative 10 % conversion efficiency.

Publically available source material contains minimal information and data regarding the scalability of Nd:YAG technology although I located one investigation of a power-scalable diode laser-pumped CW Nd:YAG laser. It required a chain of injection-locked slave lasers to achieve an efficient, frequency-stable, diffraction limited beam.[22]

Another report claimed an achieved 100 kW output from an Nd:YAG laser intended for unspecified terrestrial military applications where the range would be limited to a '*few kilometers*'. Further investigation of this claim revealed a separate paper describing a total of 7 x 15 kW Nd:YAG lasers coherently combined with a master oscillator power amplifier (MOPA) to achieve the ~ 100 kW output with a beam quality unspecified but described as '*good*'.[23] In 2012 the US company Raytheon demonstrated a claimed '*solid state fibre laser*' producing a 50 kW beam able to engage unarmoured and slow-flying UAVs. It was unclear from information provided if the device comprised a single laser or more than one lower-powered lasers functioning cooperatively in some manner.

Research into higher-powered Nd:YAG lasers is continuing commercially, and conceivably within classified military installations as well. There is, however, a possibility that Nd:YAG and similar lasers will never reach the minimum output power required to catastrophically damage a targeted spacecraft because, '*A principal obstacle to scale-up to megawatt power levels is the removal of waste heat from the solid state medium. As the gain medium heats, optical quality is lost as the medium distorts and eventually heating will kill the gain.*'[24]

Assuming Nd:YAG lasers could be coherently combined external to a laboratory setting to apply 100 kW output, then a potentially significantly destructive beam could be projected with a coverage of 2 cm^2 (assuming the 50 kJ/cm^2/s guideline applies). The required power input, based on 10 per cent whole-of-system conversion, would then be

~1 000 kW. However, the challenge of containing the beam to an area of 2 cm^2 implies application over short distances only and endorses the claim cited above that the terrestrial laser weapon would have had an effective range of only a few kilometres.

From information and data located it seemed improbable any weapon based on diode-pumped lasing and capable of inflicting significant damage to an orbiting object could exist in c.2014 unless additional pulsing and amplification technologies could be utilised. The practice of coherently combining these lasers to achieve a relatively high output implies other approaches to amplification have not produced the required outcomes. Technological challenges including the in-space power requirement might be resoluble. However, progress achieved over past decades implies any such weapon designed to attack spacecraft will probably not become available before 2030.

The wavelength of the laser energy from free-electron lasers (FELs) can be changed by modifying the properties of the laser's integral magnetic field, thus FELs offer significant flexibility in their operation.

The FEL military laser, as perceived in 1985, was a weapon with an estimated 120 m length, not including its power source. It was then considered a candidate terrestrial weapon capable of being fired *into* outer space. Progress has been made in addressing the physical dimension challenge, although to launch such a device might not be feasible unless on-orbit assembly is proposed. A study completed in 2008 concluded '...[recent advances] *suggest that the necessary optical cavity could be contained within a 20-meter-long structure*'.[25]

One study was located which included the power to output ratio of a FEL. In that instance, a primary input of 13 400 kW was required to produce an output of 3 200 kW an efficiency ratio of approximately 24 %.[26] Applying the 50 kJ/cm^2/s criterion for inflicting damage on most spacecraft, the laser could support a beam of that intensity covering 64 cm^2 which indicates potential for application as a potent weapon in outer space.

However, another study considered a hypothetical 1 000 kW FEL and identified an input power requirement of 20 000 kW, indicating only 5 % efficiency.[27] An additional and integral consideration was that the terrestrial 3 200 kW FEL, referred to above, converted 9 700 kW of its input power into heat.

During 2006 the US Navy, which had maintained an interest in FEL development, reported an experiment which achieved 14 kW with a wavelength of 1.6 μm. The announcement was similar to another claim, released on 3 May 2007 by the US Department of Energy-owned Jefferson Laboratory, that it had achieved a 14.2 kW output at the same wavelength, leading to the possibility these were references to the same experiment. A website operated by the US Office of Naval Research included a fact sheet, *Free Electron Laser*, confirming ongoing interest in the technology but it was ambivalent as to the applications intended.[28]

On 19 March 2010 Boeing Corporation announced it had completed the preliminary design for a ship-borne FEL. And on 14 September 2010 *Space News* announced Boeing Corporation had been awarded a contract to design a FEL capable of 100 kW output. However, during a hearing in June 2011, the US Senate Armed Services Committee recommended termination of the project. At that time the ship-borne laser was revealed as having been planned to be available in 2020.[29]

Available evidence implied FEL technology in 2014 was at best verging on practicability for deployment as an anti-satellite weapon. Considering the flexibility of FELs and the potential to supply power via a collocated nuclear reactor, it is likely research on them for military purposes will have continued in the US and therefore probably elsewhere. In the foreseeable future FEL technology could be developed for deployment as a viable anti-satellite weapon. Progress made with the design of nuclear reactors for potential deployment into space implies an adequate power source could become available by 2030. However, the challenges of size and mass, together with an anticipated strong thermal signature, might confine such a weapon to terrestrial installations from which it would fire into outer space.

One further indication of the high-mass characteristic of weapons-grade lasers is that published images of US terrestrial lasers have commonly displayed these devices mounted upon all-terrain trucks. The trucks appeared to be the Heavy Expanded Mobility Tactical Trucks manufactured by Oshkosh Defence. Specifications for a variant of the vehicle were found and that version had a combined truck and trailer payload of 33 tonnes.[30] Information regarding the category of laser allegedly depicted was not found. Other published images have shown lasers of unspecified category and capacity ostensibly operating from US Navy vessels in the destroyer class, commonly in the displacement range of 8 000–9 000 tonnes, on which 30 to 100 tonnes of laser equipment could be accommodated together with fuel sufficient for power and time-limited experiments.

Concerning developments of high-powered lasers in States other than the US, one study of published papers and other sources concluded the PRC had an active program to develop lasers as directed energy weapons. [31] PRC research was found to include COILs and FELs while research related to FELs overlapped with non-nuclear X-ray technology.

A research program was also continuing into advanced technologies supporting laser applications including deformable mirrors and micromachined membranes. These observations support other assessments that the PRC could attempt to develop a laser-enabled space weapon at an unspecified future time. Additional evaluations have reached more dramatic conclusions but failed to provide supporting evidence or verifiable and independent references.

A 2012 report from Russia claimed it had advanced beyond the US in developing some laser weapons. Other analyses of Russian progress have been less enthusiastic. A Gas-Dynamic Laser, the Almaz-Antey High Energy Laser Directed Weapon, has apparently been used by Russia as a test bed to accumulate expertise across a range of laser applications.[32] Illustrations included in the report showed a series of apparently integrated heavy-duty trucks carrying components of the laser technology thus implying significant mass.

An airborne laser of uncertain power has nevertheless been flown on a Russian aircraft. Russia began an airborne laser weapons program, impliedly in the 1980s. The weapon was then designed to be carried by a Beriev A-60 aircraft. The former USSR was using at least two and possibly three such aircraft for laser tests against high-altitude balloons in 1983 – 84. Russia might, in 2009, have resurrected an earlier laser program.[33]

In common with other States, France has adopted laser technologies for a range of military applications. The Thales Company has been reported as working cooperatively with France's Ecole Polytechnique to develop a directed energy weapon for ground and naval use. No evidence of France developing a space-based or space-directed laser was located.

One German company, Rheinmetall, has claimed successful development of a powerful laser weapon. Further investigation revealed its powerful weapon to reportedly comprise two 5.0 kW laser modules plus a 1.0 kW laser, those figures having been provided without further elaboration. The category of laser technology employed by Rheinmetall was not disclosed. A panel discussion at the 2012 Directed Energy Systems Conference, Munich, concluded it was likely to be 10 to 20 years before laser generating devices were widely deployed as (impliedly terrestrial) weapons.[34]

The on-line report *Space War* recorded in 2007 the Republic of Korea (RoK) was developing a mobile, truck-mounted laser capable of destroying DPRK Missiles. No details of proposed power or laser category were revealed. No subsequent references to this alleged development, or any confirmation of it, were found.

According to a report dated August 2010 India was developing a laser-based anti-missile system. The Indian laser was limited to a range of 10 km although scientists were claimed to be developing a weapon with an 80 km range. [35] It was not clear from information available whether India was attempting to develop a laser capable of being deployed in space or one limited to applications as a terrestrial weapon.

For at least 20 years Israel has been cooperating with the US to develop tactical laser weapons.[36] Israel has worked with the US corporation Northrop Grumman to develop a Tactical High Energy Laser in both fixed and mobile versions. That laser is based on deuterium-fluoride fuel and intended to defend against tactical missiles and artillery shells. I did not find estimates of its lasing power.[37] On 19 January 2014 an Israeli media outlet announced Israeli would in 2015 deploy a domestically manufactured laser system, the '*iron beam*' with a range up to 7.0 km against short range rockets and mortars. I did not find evidence that Israel had been attempting to develop a laser weapon for deployment into outer space.

In 2012, Iran claimed to have built a laser weapon for terrestrial applications but provided no details, rendering impossible even the most generalised analysis of that assertion. Russia apparently became aware of the claim in advance of it being made and dismissed it as '*fiction*'.[38]

In 2012 a paper was published claiming to have confirmed the potential to perturb space-debris orbits using photon pressure from ground-based lasers, thus reducing the potential for collisions in outer space.[39] Whether this technology could be scaled-up or modified to have additional applications in outer space is not yet apparent.

One parallel development involving laser technology in the US, France, the PRC and Russia has been research activity associated with using laser power to initiate fusion for domestic-terrestrial purposes. The US has one facility dedicated to this work, the National Ignition Facility, but France, the PRC and Russia are building related new capabilities (France, *Le laser Mégajoule*; PRC, Divine Light 4; and Russia, Experimental Physics Research Unit). Some aspects of the related science have clearly proliferated beyond the four States mentioned. In 2012 Japan hosted a conference on Laser Inertial Fusion Energy.

While these initiatives could enhance knowledge of lasing, they are unlikely to support progress toward building a viable weapon for launch into outer space. The facilities are physically large, correspondingly immobile, and oriented toward the different outcome of providing low-polluting power and energy production for terrestrial consumption.

In summary, from information available it is conceivable, but not more, that chemical, diode-pumped or FEL lasers intended to damage spacecraft on orbit could be developed by 2030. Given the combination of challenges related to mass, size, provision of fuel and the obvious signatures of lasers as orbiting targets, terrestrial lasers capable of being fired through the dense atmosphere into outer space seem more likely to be successful than space-based lasers. A potential variation could be airborne lasers operating from within sovereign airspace, able to readily access refuelling facilities.

Particle and Related Accelerators

During the 1980s components of the US scientific community hypothesised there was potential for a '*neutral particle beam*' space weapon to accelerate particles at velocities able to catastrophically damage an intended target. In 1986 the US Office of Technology Assessment described a potential weapon in the following terms:

> *A neutral particle beam (NPB) weapon might consist of a negative ion source, a particle accelerator, beam focusing and pointing magnets, and a "stripping" device—e.g. a gas cell—which strips the negative ions of their extra electrons, thereby neutralising them, as well as a power source and other ancillary equipment ...*[40]

Early developments appeared encouraging. In 1986 hydrogen atoms from an accelerator at the US Los Alamos Meson Physics Facility were reported to have reached energies of 800 MeV.[41] However, in terms of total energy transmitted the information was almost meaningless without additional data. Reports of similar progress were recorded by in 1988 regarding an experiment by '*Los Alamos scientists*' in which laser and charged particle beams were combined to produce a steerable beam with promising results.[42]

During an experiment on 13 July 1989, within the Los Alamos-based Beam Experiment Aboard a Rocket (BEAR) program, the US produced and launched an allegedly neutral particle beam using a sub-orbital rocket as its platform.[43] The accelerator was reported as being 1.0 m in length and '*lightweight (greater than 55 kg)*', implying major and fundamental advances with both miniaturisation and mass.[44] However, when I found the final report of those tests it revealed an '*accelerator payload mass segment*' of 1 545.8 kg and a total vehicle mass of 7 111.4 kg, both of which were certainly greater than 55 kg.[45]

The device was described by its corporate author, Los Alamos National Laboratory as, '*designed to produce a 10 mA (equivalent), 1 MeV neutral hydrogen beam in 50 microsec pulses at 5 Hz.*'[46] Assuming the BEAR beam met its design specifications, energy transmitted was so slight that sensitive instrumentation would have been required to detect its presence. The fundamental challenge of mass for a potential accelerator-weapon appeared not to have been resolved at that time or with that technology.

Another high-energy experimental facility was established by the US at Kirtland Air Force Base, New Mexico, but its focus later changed to microwave experimentation. Interest in employing the charged particle technology as a potential space weapon then appeared to wane. Scaling-up from theory and demonstration modelling to the development of an effective anti-satellite weapon has continued to challenge the global scientific community. Successes have been both been both rare and fleeting.

To date, the most powerful accelerators (powerful in terms of achieving high relativistic velocities) have been physically large. The accelerator at Los Alamos Meson Physics Facility, later renamed Los Angeles Neutron Science Center, is approximately 1.0 km in length. During the 1990s the US scientific community embarked upon building an 87.0 kilometre-long research accelerator in Texas, the project being cancelled before completion. The 27.0 kilometre-long CERN Large Hadron Collider has since been constructed in Europe. A 6.3 km Tevatron at Fermilab, near Chicago, ceased operations in 2011 as a component of fiscal austerity measures. Numerous smaller accelerators and related technologies, such as linacs, synchrotrons and isochronous cyclotrons, exist globally but none of them, so far as could be determined, are candidates for launch into outer space because of their physical size, mass and because they were designed primarily as components of physics experiments requiring precise observations.

Also, projections at high relativistic velocities require impressive quantities of fuel. In 1986 US officials had estimated that projection of 10 kGy across an outer space range of 40 000 km '*might require about 25 tonnes of liquid hydrogen, liquid oxygen, tankage and "other overhead"*'.[47] Another claimed mass-to-energy-applied ratio was that 1.6 tonnes of fuel would be required at 1 000 km range, impliedly to deliver 10 kGy.[48] A further and then contemporary source estimated the mass of fuel required to cause disabling damage (otherwise unquantified) across 40 000 km as 80 tonnes.[49] In these examples the fuel-mass estimates applied to only one firing of the proposed weapon.

Regardless of how the fuel factor is calculated it seems certain that accelerators of any category designed to function as weapons in outer space will require high levels of power regardless of whether that power is derived from liquid-fuel or another medium. Indicatively, Europe's CERN Large Hadron Collider requires electrical power in the range 220 – 300 MW. The US Fermilab Tevatron, when operating, drew some 45 MW.

Despite the magnitude of these challenges, during the 1980s the US Department of Defense claimed that the USSR could, by the mid-1990s, have developed a space-based particle-beam weapon.[50] The claim was not supported by any verifiable evidence and I did not find any.

The Argonne National Laboratory nevertheless pursued this research with a Continuous Wave Deuterium Demonstrator, the projected power from which was also at a miniscule level.[51] US Department of Energy has since claimed a capacity to produce beams of protons with energies greater than 1 GeV passing through distances of '*a few kilometers*' without appreciable losses. The devices were apparently designed for terrestrial inspection roles rather than as precursors to destructive applications.[52]

In 2010 the US Department of Energy again highlighted the potential for accelerators to make significant contributions to directed energy capabilities. However, the emphasis was on potential. In addition to the already identified mass and dimension problems, challenges identified included: much current data being incomplete and dating from the 1950s and 1960s, the current accelerators requiring highly specific operating environments, and '*a further challenge is to develop dedicated accelerator-based beamlines ...* '[53]

Summarising progress to 2010, the US Department of Energy asserted:

> *A clear need exists in a variety of defense and security applications for compact, robust high-gradient accelerator systems that can be deployed in the field. Currently, vibrations and temperature swings affect most accelerator-based technology, and components and infrastructure occupy a large space.*[54]

In 2013 the University of Texas announced success in building a '*tablet*op' particle accelerator able to accelerate '*about half a billion electrons*' to 2 GeV, but the physical projection range was '*about 1 inch*' [~ 2 cm].

Minimal progress appears to have been made by the US over the past 30 years with challenges obvious in the 1980s still unresolved in 2014 and conceivably not resolvable before 2030 CE and likely far beyond that.

Nevertheless, in 2011, using language similar to that employed during the mid-1980s regarding the USSR, US officials claimed the PRC had been developing particle beam technologies as anti-satellite weapons.[55] No further details of or justifications for this claim were located.

It is not possible to confidently confirm or repudiate many assertions regarding the PRC, given it releases minimal information regarding military research activities. However, in April 2009 the *Xinhua* newsagency recorded activation of the PRC's biggest particle accelerator, a synchrotron, in Shanghai's Zhangjiang High Technology Park. The device had allegedly cost USD 176 million, a restrained investment when compared with the price of some established accelerators and a possible indication that development of high-powered particle beam weapon might not have been a significant priority for the PRC. [56]

Additionally, a *The New York Times* report from 2011 recorded the meeting of an international consortium in Beijing to plan a new particle accelerator costing an estimated USD 6.7 billion with a length of 31 km extendable to 50 km.[57] This report implied that Chinese interest in further developing particle acceleration technology had persisted. However, with the internationally funded and staffed facility proposed, the research instrument, if it is built, would not seem compatible with the clandestine development of a new weapon for deployment into or through outer space, although some data obtained might be used in that way.

Knowledge of accelerator technology is now ubiquitous. The European Laboratory for Particle Physics involves participation by most European States. Russia, India and the US are among the States with observer status, and cooperative agreements have been reached with additional States including Iran, Australia and the PRC. Russia has additionally enjoyed discrete access to accelerator technology through its scientific facilities at Dubna near Moscow, and Protvino near Serpukhov. In 2010 the US Department of Energy estimated there were some 30 000 particle accelerators world-wide. Many had been designed for and were limited to medical or industrial applications.

I found no literature suggesting France, any other European State, Japan or Israel had committed to building a particle beam weapon. The technology to build, launch and operate such a device in space seemed beyond the conceivable reach of Iran, Pakistan or the DPRK.

Perhaps unexpectedly, India has developed a device named the Kilo Ampere Linear Injector (KALI). This is a relatively new and difficult-to-categorise development. KALI has been described (by a pro-Pakistani source) as a form of linear accelerator which, it was claimed, could be adjusted to produce X-ray or microwave frequencies. It allegedly has some potential for eventual development into a weapon.[58]

India has acknowledged possession of the device adding that a recent version, the *KALI 10 000*, had a mass of 26 tonnes, was '*very power hungry*', had a long but unspecified recharge time and required a cooling tank of 12 000 litres of oil.[59] The device appears to be at an early stage of development and, in common with other broadly similar technologies, it might resist scaling to high output levels.

Given the enviable credentials of US research institutes involved with particle beam design over past decades, and the apparent lack of progress made, the likelihood of any other State orbiting a viable particle beam weapon capable of damaging space vehicles, by 2030, seems remote. The US appears to be moving toward an acceptance of this conclusion in relation to its own research.[60]

Overall, there appears to be no credible likelihood of any weapons-grade beamed-energy devices, based on either lasers or other technologies, emerging within a foreseeable timeframe for use against spacecraft, with the possible exception of terrestrially-based devices of limited utility. Propagandistic claims, unsupported by verifiable evidence, will likely persistent with attempts to create the opposite impression.

Chapter 5. Other Technologies of Space Warfare

In this chapter I use the term *disabling* technologies to discriminate between technologies intended to inhibit the performance of spacecraft from those, previously considered, designed to destroy spacecraft or cause catastrophic physical damage to them.

These techniques would nominally include interfering with electronic systems, as well as using various forms of camouflage or deception technologies, possibly in combination. India, Japan, Israel and some European States, notably France, Germany, Britain and Italy, appear to have the advanced electronics and other skills needed for developments of this nature. Occasional hints have emerged that India is developing disabling technologies to use against spacecraft. A US paper in 1986 claimed, with regard to high-powered microwave research, '*France, Germany, Switzerland and Japan are also wasting no time in following suit in this area.*'[1]

Radiofrequency and Microwave Weapons

Cyber warfare is beyond the defined scope of this work for reasons I gave in the *Introduction*. Radiofrequency-applications can however be involved more directly in both the attack and the defence functions of space warfare.

An example of radio frequency transmissions being used to jam a satellite occurred in 1996. The APT Satellite Company leased from Tonga a disputed position in the geosynchronous orbit. APT manoeuvred a satellite, *APSTAR-1A,* into the leased location, it having already been occupied by another satellite, the *Palpa B1*. *Palpa B1* was owned by the Russian Gorizont organisation but leased to a third party. In response to the APT initiative, *Palpa B1* was set to transmit so it interfered with *APSTAR-1A* until the APT Company was able to resolve the problem by adopting a technological solution to ensure its satellite did not inhibit the operation of *Palpa B1*.[2] Another example of apparently deliberate jamming originated in Cuba, c.2003, and was made against a national Iranian television station operated by an Iranian dissident group located in Los Angeles.

The US has developed electronic warfare systems designed to disrupt Earth-to-space, space-to-Earth and space-to-space communications.[3] The US in 1995 deployed a number of jamming devices, apparently referred to as, or code-named, 'Wolf'. By 1994 the US Air Force had apparently created an organisation, 76th Space Control Squadron, specifically to use terrestrial jamming stations to disable or jam foreign satellites.[4] The claim is congruent with other announced US intentions and demonstrated capabilities.

Developments of this nature when known or suspected can lead to international tension. In November 2011 a Russian spacecraft, the *Phobos-Grunt*, was launched to travel to Mars but became unexpectedly confined to a low Earth-centric orbit. Accusations then surfaced in Russia that the problem had been caused by emissions from the US High Frequency Active Auroral Research Program located in Alaska, one such claim being attributed to Russian Lieutenant-General Nikolay Rodionov.

Information regarding USSR-Russian anti-satellite jamming activities is sparse. However, there exists significant published evidence of comprehensive *terrestrial* radiofrequency jamming undertaken by the USSR and there is no obvious reason why these technologies could not be employed in outer space. Given the US's well publicised initiatives it would be disingenuous to believe the USSR-Russia had not followed suit or even taken the initiative in this activity.

A similar situation applies to the PRC which has a record of jamming terrestrial radiofrequency signals.[5] The US has also accused the PRC of actively jamming radiofrequency signals from satellites. A report citing the PRC news outlet *Xinhua* described the launch of a PRC '*anti-jamming satellite*' implying Chinese awareness of anti-satellite jamming activity in outer space.[6]

Iran is also alleged to have jammed satellite signals, suggesting activity of this nature can be pursued by States other than the main space-faring nations.[7]

The challenge of interference with satellite transmissions has now reached the level where several specialised organisations have been formed to advise operators on reduction or avoidance techniques.

Additionally, claims of radiofrequency devices being deployed intentionally against timing or positioning signals transmitted through outer space by the NAVSTAR Global Positioning System (GPS) are common. These assertions are seldom supported by evidence although classified records of such activities might exist. It is however well known that terrestrially-based jammers can block or distort GPS signals within a few kilometre range of the jammer.[8] These jammers are not space weapons; however, given the effectiveness of terrestrial jammers, there is no doubt it would be technically feasible to orbit a jamming device aimed at a global navigation satellite system.

The protection of Global Navigation Satellite Systems (GNSS) satellites from jamming is constantly under review and is being improved. Moreover, any such attack could invite reciprocation or other reprisals, conceivably aimed at similar capabilities. In this context navigation and related satellite systems, in addition to the US-owned GPS, are either already in operation or being developed by States with significant military capabilities. They include: Russia (Globalnaya Navigatsionnay Sistema (GLONASS)), the PRC (Beidou), Japan (Quasi-Zenith Satellite System (QZSS)), France (Doppler Orbitography and Radiopositioning by Satellite System (DORIS)), India (Indian Regional Navigation Satellite System (IRNSS)), and the European Union (Galileo).

There exists a distantly related concept of employing electromagnetic energy at microwave frequencies as a weapon to attack objects in outer space. The designed device would transmit energy at microwave frequencies to enter a spacecraft and disrupt, or possibly permanently damage, electronic components. The attacks are commonly characterised as 'front-door '(entry via satellite antennae) or 'back-door' (entry by any other means). An effective attack requires some knowledge of the target spacecraft's technical specifications as well as an ability to circumvent or negate any overload or related protective devices installed in it. Consequently, the effectiveness of a microwave attack will commonly be less certain than the outcome of an attack using a kinetic or nuclear weapon. The likelihood of an effective back-door attack can be increased by using a spread of microwave frequencies in the hope of finding an exploitable vulnerability.

The US appears to have studied the potential to employ microwave amplification by the simulated emission of radiation (MASER) since the early 1960s. A summary of this and other US microwave energy initiatives from the 1980s has been complied by the University of Bradford.[9] One such project was designated Project Blackeye. The technology used might have had a capacity to disable some satellite sensors.

In 2009 the USAF was reported to be engaged in a program to develop a missile able to pass close to more than one target while emitting microwave energy sufficient to damage electronic components. The program was named Counter-Electronics High-

Powered Microwave Advanced Missile Project. Obtaining an adequate and portable power supply was identified as a challenge to be addressed.

In March 2011 The USAF awarded to Lockheed Martin Corporation a contract to produce a concept for a microwave weapon, conceivably as part of the program mentioned.[10] From the information and data then available it was inferred, but not more, the device could be used in outer space. In outer space the principle technical issues limiting the application of microwave weapons have been acquiring the enabling power levels and marrying the emitter to a suitable platform.[11] Microwave-frequency weapons at low power settings might also have a radiofrequency jamming function against some transmissions.

Microwave technology continues to be developed for disruptive and possibly also for destructive applications. In 2011 the US was also reported as having built the Active Electronically Scanned Array, a radar able to be converted so as to rapidly focus microwave energy into a concentrated and potentially damaging beam.[12] The US reportedly had discovered ways to employ microwave frequencies able to circumvent some conventional anti-jamming methodologies such as Faraday cages.

Microwave-weapons technology is not confined to the US. Its underlying principles are well-known internationally. Russia has developed the Ranets E microwave system with an average output power equivalent of 2.5 – 5.0 kW. It is mounted on a large truck, implying mass would be one factor constraining launch into outer space.[13] India has claimed to be developing a high-powered microwave for use as a futuristic weapon.[14] Closely related developments in the PRC have been reported over recent decades.

However, the technology continues to attract controversy. A 2012 paper in the journal *Nature* concluded, after some 50 years of attempts, no viable microwave weapon had yet been produced and cast doubt on whether such an outcome would or could emerge.[15] Conversely, because so many States are dedicating resources to the task it seems premature to assume a viable anti-satellite microwave weapon will not be developed within the next two decades.

Space Planes and Hypervelocity Vehicles

The concept of vehicles able to both fly and orbit, thus functioning in both the terrestrial and outer space regions is well-established. To date, these vehicles have at best illustrated a capacity to launch from Earth into outer space, perform limited manoeuvres in outer space, including docking with another space vehicle, before returning to Earth and being landed either by a human pilot or autonomously.

The development of space planes extends back to at least 1951 when the US began planning for construction of an extreme-altitude craft, later designated X-15. By late 1961 evolving variants of the vehicle had reached altitudes '*in excess of 200 000 feet*' (> ~ 61 km) and in 1963 US pilot J Walker flew one version to an altitude of 354 200 feet (~108 km).[16]

A US Space Shuttle was first launched in 1981. This populated vehicle was designed to travel in orbits between '*about 115 to 400 miles high*' (~ 185 to 644 km) before returning and landing in a manner similar to a conventional aircraft.[17] In 1984 the US organisation DCS Corporation submitted under contract to Defense Advanced Research Projects Agency a '*preliminary analysis of the potential for a small, new, generic type of manned spaceplane as a manned research vehicle*'. A 1986 report by the RAND research organisation referred to a National Aerospace Plane Project.[18]

In his State of the Union address delivered on 4 February 1986 US President Ronald Reagan raised the a possibility of a vehicle able to fly in the atmosphere or in space, stating;

> '*... we are going forward with our research on a new Orient Express that could, by the end of the next decade, take off from Dulles Airport, accelerate up to 25 times the speed of sound, attaining low earth orbit or flying to Tokyo within two hours.*'

The US tested an Experimental Spacecraft System-11 (XSS-11) in space during 2005. Subsequent tests allegedly involved close inspection visits to satellites. The US Air Force had also been engaged in a project known as the Common Aero Vehicle which in c.2005 remained in the research phase with minimal information apparently migrating to the public record. An 0.85-scale version known as the *X040* might have been tested between August 1998 and July 2001.

The Commander of US Air Force Space Command has been quoted as informing the US House Armed Services Committee in 2005:

> '*This [Common Aero Vehicle] is an incredible capability to provide the warfighter with a global reach capability against high payoff targets ... This is the type of Prompt Global Strike I have identified as a top priority for our space and missile force.*'[19]

As at late 2013 the US Defense Advanced Research Projects Agency was claimed to be managing the Experimental Spaceplane (XS-1) program, the technical objectives for which included flying 10 times in 10 days. The US had also built and tested in outer space another space plane designated *X-37B*. US Air Force Colonel Andre Lovett, Deputy Commander of 45th Space Wing, was quoted as stating '*This* [the 22 April 2010 launch of an X-37B space plane] *launch will help ensure that our warfighters will be provided with the capabilities they need in the future.*'[20]

Space plane development has not been confined to the US although, so far as I could determine, as at late-2014 the US was technologically more advanced than other States in the development of a space plane with potential military applications. However, space planes appear to be within the capacity of States experienced in the combination of space launch, advanced aircraft manufacture and remote control technologies such as are employed with UAVs. These States would include, in addition to the US, the PRC , India, France, Japan, Israel, and Russia. Accelerated development of space planes could be enabled by indigenous innovation, technology exchanges, reverse engineering and data theft.

PRC space plane development apparently began in the late 1970s when Qian Xuesen proposed a craft similar to a conceptual US vehicle known as Dyna-Soar or X 20. The PRC has since been investing in space plane technology through the China Academy of Launch Vehicle Technology. During c. 2009 the PRC was testing a vehicle named *Shenlong* (loaded mass estimated as 5.1 to 7.5 tonnes) to validate space plane technologies although PRC ambitions to build a vehicle with a launch mass of 100 tonnes have been recorded. One source has quoted Commander General Xu Qilang as discussing '*air and space integration, possessing capabilities for both offensive and defensive operations.*'[21]

Reports have referred to the development of an Indian space plane, the Aerobic Vehicle for Hypersonic Aerospace Transportation (AVATAR).[22] However this project was not listed by the official India Space Research Organisation website and its status is unclear.

In the early 1980s the French organisation Dassault began development of a space plane with either the vehicle or the entire project named Hermes; however, the project was cancelled in 1993. In December 2011 the European Space Agency was reported as preparing to launch a space plane, the *IXV.* In December 2012 the Institute of Space Systems at the German Aerospace Centre was conducting a feasibility study for a 50-passenger hypervelocity vehicle.

During the 1990s Japan was developing a re-entry vehicle, the *HYFLEX.* The program was abandoned but might have provided data to support a space plane. Also in the 1990s Japan began developing a form of space plane referred to as the *HOPE-X.* This project also was cancelled or abandoned before a launch had been achieved and references to it were not found in literature after 2003. Subsequently, Japan built the Kounotori HTV-3 spacecraft for autonomous deliveries of cargo to the ISS, although it was designed to self-destruct upon re-entry into the dense atmosphere.[23]

In January 2012, a computer terminal connected to the Japanese Aerospace Exploration Agency was infected by a Trojan virus. This led to suspicions that technical data relating to the HTV had been stolen, opening the possibility that aspects of Japan's intellectual property regarding space plane technology would proliferate.[24]

The USSR built a space plane, the Buran, which was similar in external appearance to the US Space Shuttle. The vehicle was launched into outer space, once, on 15 November 1988. It was unmanned and after two orbits was landed autonomously at Tyuratumthus. The mission demonstrated the USSR had acquired, and Russia has almost certainly inherited, the necessary technological skills to build more advanced space planes in the future. During 2011 a rumour emerged that Russia was building, or considering building, a modern space plane but I could find no confirmatory data or information.

Although some space plane technology has been developed ostensibly to re-supply the ISS there is no doubt that evolved vehicles would have potential military applications. An advanced space plane could move in and out of orbit and deliver weapons to strike almost any populated location on Earth. Such a vehicle could also launch, service, repair or recapture either satellites or co-orbital weapons with minimal publicity. The US Space Shuttles, for example, routinely launched and recovered satellites as can Russia's ISS resupply vehicle.[25]

Given their demonstrated docking capabilities, some space planes could be used for close inspections of orbiting spacecraft to determine their design characteristics and missions. A space plane equipped with a robotic arm, in principle similar to the mechanical arm attached to the ISS, could physically damage satellites.

The space warfare implications of corporate projects, such as SpaceShipOne or Virgin Galactic, remain unclear. Compared against the possibly long-duration requirements of a military space plane, these vehicles involved appear to be fragile machines designed to transport space tourists under benign conditions. However, the financial economies associated with their design and construction might provide precedents for a new generation of military space planes, and they offer potential to launch space-related sub-orbital experiments such as the BEAR trial mentioned previously.

Regarding manoeuvrable space vehicles generally, some observers have claimed the penalties of additional fuel and increases in launched mass required to enable each manoeuvre in outer space are so great that the process would not be financially economical. Additionally, the mass of fuel needed for orbital manoeuvres would either limit the flexibility of space planes or reduce their potential payload capacities. The contrary perspective is that financial costs and related penalties have seldom constrained military ambition, particularly if a strategic benefit was perceived.[26] From records available and cited there seems little doubt space planes are becoming a new form of potential space weapon, and impliedly a new category of target, in outer space. Already, the number of space planes existing or under development is in excess of requirements to re-supply the ISS or its PRC equivalent, *Tiangong-1,* or its planned successors.

One significant implication of evolving space plane technology was not mentioned in any literature found. That implication is the potential to apply space-plane engine and related technologies to missiles. This would allow missiles to be launched, stop their engines, enter ballistic trajectories, restart their engines, change orbits, and repeat these manoeuvres several times if necessary. Such capabilities make it fundamentally more difficult for any outer-space weapon, or terrestrial anti-missile defence system, to acquire and attack such missiles in outer space. The technology would also offer any such missile an expanded menu of targets either in outer space or terrestrially. It is a rational, achievable and obvious extension of the space plane-engine technology.

Hypervelocity vehicles are sometimes considered in company with space planes, although the hypervelocity technologies are profoundly different to those used for propelling craft of any description into or through outer space. Literature concerning hypervelocity vehicles has focussed on vehicles propelled by scramjet engines.

The scramjets are confined to high atmospheric altitudes and normally operate in the altitude range ~28 – 34 km, making them terrestrial systems and not space vehicles. They are mentioned here because they have been designed to operate above altitudes commonly achieved by aircraft. They could, when or if further evolved, affect activities in outer space by, for example, launching weapons designed to affect spacecraft.

The US has been developing at least two scramjet-powered vehicles. One, the *Hypersonic Technology Vehicle-2*, was tested on 22 April 2010 and again on 11 August 2011. Details of its performance specifications were not released although the US Defense Advanced Research Projects Agency has claimed a designed (but conceivably not achieved) top speed of Mach 20. An apparently separate vehicle, the *Advanced Hypersonic Weapon* with a designed range of 3 700 km was tested on 18 November 2011.

Russia and Japan both claim to have developed the scramjet motors integral to powering hypervelocity vehicles.[27] The PRC and India also have made some progress toward demonstrating this technology. [28] The PRC is recorded as developing scramjet technology at its Shenyang Aircraft Design Institute (Institute 601) and its Chengde Aircraft Design Institute (Institute 611). It was reported as having tested a 'hypersonic missile' on 9 January 2014.[29]

An Australian-developed scramjet motor was built but a field test at the Andøya Rocket Range, Norway, ended when the '*payload did not reach the correct conditions to begin collecting data as planned,...*' viz. the launch rocket crashed.

Knowledge of scramjet principles is thus widespread but there will be variable technological and chronological gaps between States comprehending the scramjet concept and then attempting to build a working model.

Unknown factors affecting applications of scramjets include whether scramjets will prove scalable to the level where they have military applications and whether the technology will prove practicable given each scramjet requires a launch vehicle to carry it to an altitude from which it can attain the momentum needed for its engine to begin functioning.

Given the extent of interest in hypervelocity travel, some form of incorporation into technologies with implications for warfare in outer space seems possible before 2030.

Battle Stations in Orbit

Battle stations have been proposed for insertion into orbit although available literature is imprecise regarding the definition of a battle station. It is inferentially apparent the concept incorporates an entity hosting either one or several different categories of space weapons to enable both self-protection and offensive engagement.

Given conclusions already drawn regarding the potential physical dimensions of high-powered weapons in space, a battle station would of necessity be large and require assembly in space from launched components. In addition it would need to be highly reliable, sufficiently agile to survive attacks from the range of feasible space weapons, yet still function and remain able to achieve precise target identification, tracking and weapons delivery, as well as remaining interactive with a command and control architecture and responsive to a potentially changing threat menu. Progressive technological obsolescence would also present challenges.

To remain potentially responsive to most events occurring in near-Earth space, at least three such devices would be required in GEO and they could need supplementation by other weapons systems to provide coverage of the high-latitude polar regions. Refuelling or resupply to GEO would be challenging, given the reduced payloads able to be conveyed to that altitude.

On 15 May 1987 the USSR apparently attempted to launch a single battle-station, or a component of one, known as the *Skif*, but the launch failed. Another version of the same event has claimed *Skif* was a demonstration version of a space-based laser.[30] No other attempt by any State to launch a battle station, or any technology resembling one, was located. It seems unlikely a large and obvious potential target, such as the battle station envisioned in the 1980s, will be launched into outer space during the foreseeable future.

Other Relevant Technologies

A large and increasing number of additional techniques and technologies have been or could be utilised to affect the conduct of warfare in outer space. Following is a selection of examples, some likely to be more effective than others.

Reflecting long-established techniques employed in terrestrial warfare, attempts to camouflage satellites have been made and apparently are continuing, although contemporary details seldom reach the unclassified environment. I found a 1994 patent for a shield to be deployed after launching, and designed to camouflage the physical characteristics of a satellite.[31] Techniques for avoiding radio frequency interference include reductions in the detectable emissions from space vehicles, a concept often referred to as '*achieving lower probability of interception*'. Space vehicles can be coated with substances which absorb or fail to accurately reflect the signals of interrogating radars so creating a lower probability of detection.[32] [33] On 3 June 2011 the Head, US Strategic Command during an oversight hearing by the Senate Armed Services Committee, Strategic Forces Panel, declared that some satellites '*aren't what they seem*', having been camouflaged to appear as communications satellites.[34]

These defensive techniques often require an adversary to use radars and related devices at peak power in an attempt to overcome spacecraft defences. This in turn reveals the location of the adversary's radars thus exposing them to attack.

The use of decoys in outer space is well-known and established. Decoys can be equipped with the capacity to transmit misleading signals and thus distract space weapons away from potentially destructive missiles or other militarily more valuable spacecraft. These deceptive technologies, when undetected, further reduce the potential effectiveness of some weapons in the outer space region.

Allegedly, natural and environmental disturbances could, theoretically, be used to camouflage a disabling or destructive attack against a space vehicle in outer space. However, natural events are an imperfect disguise for human intervention.

High-energy electrons associated with the solar minimum seem likely to have caused damage once, in January 1994, to three satellites.[35] On 24 October 2011 a coronal mass ejection compressed the Earth's magnetosphere to the point where a concentration of solar wind particles penetrated GEO, theoretically offering an opportunity to camouflage some forms of electronic attack against satellites in that orbit. But the penetration persisted for only five minutes.

To be regarded as a credible and natural event, an attack would have to be against a spacecraft located within the arc of its orbit that was exposed to solar or other energy of sufficient magnitude to cause natural damage at the time of an attack. Energy delivered by the attack might have to penetrate a naturally turbulent electronic environment with sufficient force, integrity and accuracy to reach and damage the designated target spacecraft. The weapon able to deliver that energy would have to be available, within range of the target spacecraft with its aggressive operation remaining undetected by sensors available to an adversary.

A claim that peak solar flux temporarily blocked signals transmitted by GPS satellites, or satellites in any similar constellation, is unlikely to provide a convincing camouflage for contrived electronic interference. Terrestrial GNSS receivers did lose signal lock for a few minutes on both 6 and 13 December 2006, but the forecast frequency of such events is they will occur on average once per decade, with readily observable interference of lesser magnitude occurring on average approximately only five times per decade.[36]

The 13 major meteoroid events affecting planet Earth each year result in almost continuous exposure of spacecraft to a possible impact. However, both the mass and the spatial densities of meteoroids approaching Earth vary significantly, as does the corresponding potential for a collision with a spacecraft. The danger to any spacecraft from collision with meteoroids will be governed by the physical dimensions the spacecraft presents to incoming meteoroids, the relative trajectories of the spacecraft and the meteoroids, the spatial and mass densities of the meteoroid shower and the duration of the spacecraft's presence in outer space. Catastrophic damage to a spacecraft from this source is conceivable although historically improbable. Kennewell and Vo, 2013, estimated an expected loss of one spacecraft every 20 years from this cause.[37]

Attacks against spacecraft during meteoroid events would almost certainly attract suspicion and possibly retaliation if other satellites in the same region were unharmed, if strategically important satellites belonging to only one State were damaged, and especially if significant satellites belonging to a perceived aggressor had coincidentally survived intact.

Taking account of these considerations, camouflaging a deliberate kinetic attack as a meteoroid impact would be challenging although not impossible. Paradoxically, a genuine meteoroid collision with a spacecraft might be misinterpreted as an attack during a period of high international tension.

In the paragraphs following I have considered very small satellites and co-orbiting satellites together for two reasons; some small satellites could be used as co-orbiting spacecraft and, given limitations to the functional value of microsatellites, considered below, it is possible if not probable that some research ostensibly into small satellites has been undertaken to disguise the development of co-orbiters.

Small satellites, or microsatellites, have sometimes been proposed as substitutes for larger satellites damaged during conflict in outer space.[38] However, the value of microsatellites in these roles might be very limited.

Microsatellites in the communications role would, when in the lower orbits, appear to rise and set in relation to a terrestrial observer or in relation to a fixed ground station. Efficient communication with such satellites would desirably incorporate gimballed antennae able to track the satellites for the time they were within view. Continuous access to communications satellites positioned in low orbits would require a dense constellation of satellites, similar to the Iridium architecture, and this is unlikely to be achievable by a rapid response arrangement.

For other categories of space-based activity, miniaturisation of satellite-borne sensors would involve trade-offs as to the quantity, quality and therefore the utility of data collected. These limitations could be reduced in the case of some sensors by use of stereoscopic and related observation techniques. However, those techniques require yet more satellites which in turn add to financial cost and create additional demands for rapid-launch support.

To be effective, a program of rapid replacement in the face of satellite losses caused by conflict would need to have available a diverse range of sensors. This inventory of satellite sensors, possibly with some being duplicated, would need to be kept ready for almost immediate launch. The sensors would require compatible ground stations and data processing equipment. To collect data from these recently-launched satellites, terrestrial receiving equipment would need to be prepositioned within their downlink communications footprints. Alternatively, additional relay satellites would be required in higher orbits to receive and retransmit the data collected.

Sensors designed for carriage by a microsatellite would have to be physically compatible with a microsatellite bus (viz. the technological housekeeping component of a satellite) via a universal union architecture. Alternatively, satellite buses already integrated with sensors could be kept available but at significantly higher financial cost. Yet more sensors and satellites, however configured, would be required to allow for launch failures, subsequent destruction or technical failure on-orbit.

To support these small spacecraft, geographically distributed handling equipment and operating skills could need to include a capacity to manage dangerous chemicals or unstable and volatile fuels associated with satellites and their sensors. The fuels would predictably include monopropellant chemicals as well as the cryogenic fluids needed to support synthetic aperture radar sensors. Launch vehicles and launch platforms in sufficient numbers, fixed, mobile or airborne, would have to remain available.

There would be little point to launching replacement satellites into an environment where either extensive pollution from orbiting debris or from residual NDRE was already destroying existing space assets. Under those circumstances the replacement spacecraft would be at risk from premature termination by the same cause and could, if destroyed, contribute to an extant pollution problem.

Given the range and complexity of prerequisites for using microsatellites as substitutes for battle-damaged space assets, or as sources of supplementary capability, the future of such technology appears limited both now and into the indefinite future, except in coorbital applications essentially unrelated to rapid replacement. More timely and less expensive responses to losses of satellites in battle could include increased redistribution of tasking to commercial and foreign spacecraft, stricter prioritisation of demands for space-sourced data and services, enhanced on-orbit protection for satellites, and the improved utilisation of terrestrial alternatives to space-sourced products.

During November 2012 there emerged an indication the USAF might be ready to accept that the notion of '*operationally responsive space*' was unsound.[39] The same challenges and disadvantages confronting the US will affect any rapid response programs adopted by other States. However, some interest in the technology persists.

One residual attraction of microsatellites is the potential to launch them clandestinely. It is already possible for small satellites to be launched from a larger space vehicle, as has been done by the US Space Shuttle, and this could make them more difficult to detect.[40] Another demonstration of this capability was the Orbiting Picosatellite Automatic Launcher program by Stanford University in which one satellite housed and then later launched six others. On 25 January 2012 Russia launched the micro-satellite *Chibis-M* from its *Progress M-13M* vehicle following a resupply mission to the ISS.

The US began its attempts to use microsatellites as co-orbital anti-satellite weapons during the 1950s. The project was code-named SAINT but it was abandoned before being comprehensively tested.[41] Details of one version of SAINT were located and it had been designed to blind satellites by releasing a cloud of paint, implying physically close proximity. Following SAINT, the small satellite or microsatellite concept, while never abandoned, apparently became less significant until resuscitated by the Brilliant Pebbles program already mentioned.

A study from 1999 recommended the US pay more attention to developing small satellites because these space assets could be employed to image, damage or destroy other satellites.[42] The USAF sponsored at least three microsatellite programs; MightySat, TechSat21, and the XSS Experimental Satellite. A USAF official has been recorded as claiming one such spacecraft could be used as a space weapon.[43]

The US Defense Advanced Research Projects Agency has also been reported as developing a micro-satellite program known as System F6. Further developmental work might have been undertaken through the US Department of Defense-funded program known as Autonomous Nanosatellite Guardian for Evaluating Local Space (ANGELS). Funding began in 2005 for development of this satellite inspection and conceivably destruction capability. Lockheed Martin Corporation was awarded a contract to develop the inspection component. The ANGELS concept might also have been designed for adaptation to electronic warfare and other applications. The US additionally developed the *XSS-10* which has closely observed the second stage of its own launch vehicle, approaching to within 30 m on several occasions.

In common with most and possibly all developments of technologies related to space weapons, the US is not the sole participant. The USSR apparently conducted trials of co-orbital satellite interceptors beginning in 1968 with spacecraft launched from the Tyuratam launch site. Convincing evidence that Russia had maintained an interest in co-orbital interceptors was not located but, given the relatively low capital cost of development it probably has done so. Russia's *Progress* rocket employed to de-orbit the *Mir* space station in 2001 was in the co-orbital category of weapons, although it could be categorised in other ways also.

Japan has developed co-orbiting spacecraft. Japanese activities have included the Micro Lab Sat program; a program intended to develop spacecraft able to find, identify, rendezvous with, image and capture a target. This might be justified as, and probably was, an attempt to develop an on-orbit maintenance capability; although the potential to apply those same technologies to achieve destructive outcomes in hostile circumstances is apparent. One such mission was named Micro OMS Light Inspection Vehicle (Micro-OLIVe). A successor program, SmartSat, included a chaser satellite (*SmartSat 1a*) and a target satellite (*SmartSat 1b*).[44]

The extent of the PRC's engagement with microsatellites has been unclear. In 2007 an observer of the PRC's military activities, Johnson-Freese, believed there was ambiguity over whether the PRC was pursuing an active anti-satellite weapons program. She nevertheless concluded the PRC had a microsatellite program and noted that, because there appeared to be a weapons element in the US anti-satellite program, the PRC might pursue this option if it determined US space-related activities were threatening PRC sovereignty. The potential for that to happen seems to exist.[45]

Subsequently, *Sing Tao* newspaper reported the PRC was at an advanced stage of developing a parasite satellite, one able to attach itself to an operating satellite. However, when this report was cited by the 2003 and 2004 editions of the US Department of Defense's *Annual Report on the Military Power of the People's Republic of China,* it generated a complaint from the Union of Concerned Scientists that the information had not been confirmed and the standard of validation applied to data provided to Congress was inadequate.[46] Claims also have been made that one PRC microsatellite, the *BX-1*, might have co-orbital and destructive capacity or potential.[47]

Notwithstanding these debates, on 3 November 2011 the PRC's *Shenzhou VIII* spacecraft successfully docked with a module of the orbiting *Tiangong* space station, confirming not only a co-orbital capacity but also implying the ability to develop a Progress-like vehicle to de-orbit spacecraft. The addition of a small warhead or grappling arm to a coorbiting vehicle would not seem technologically challenging. This might already have occurred because an unconfirmed report from 2013 described cooperative manoeuvres by three PRC satellites, *Shiyan 7*, *Chuang 3* and *Shijian 7,* one of them having a robotic arm.[48]

In September 2004 the Federation of American Scientists reported on two other satellites in the microsatellite category: Germany's *Inspector* satellite and Great Britain's *SNAP* launched in 2000 (not related to a US nuclear reactor which used the same acronym).

On 13 September 2012 the German Aerospace Centre awarded the Astrium organisation a contract to manage mission preparation for the design of a Deutsche Orbitale Servicing Mission. The objective was to develop a satellite able to refuel or repair satellites or to capture uncontrollable satellites in outer space. The technology if developed would also conceivably create a co-orbital anti-satellite weapon capability although this was not stated.

Integral to the concept of microsatellites and rapid satellite replacement or supplementation is a capacity to *launch* payloads into outer space with the minimum of delay. Some solutions to this component of the rapid response challenge are well advanced in the US. In fact, rapid launch may be the most practicable component of operationally responsive space programs.

The US possesses, for example, a rapid-launch vehicle designed to be attached beneath the wing of an aircraft and to launch satellites into low Earth orbits from a base altitude of approximately 12 000 metres. The Pegasus has now evolved through several models, since it first launched a satellite on 5 April 1990. A current version can launch a satellite payload of 450 kg mass. Pegasus has had a launch success rate of approximately 90 %.

Other States might have developed similar launch vehicles but records of those capabilities are sparse. At a presentation during 2003 Russian representatives provided their version of rapid launch, it being a converted SS-18 ICBM capable of lifting 3 700 kg to low-Earth orbit. China has in place its Kuaizhou operationally responsive space program featuring a launch vehicle. I found few details of the vehicle other than claims it was solid-fuelled and could launch 400 kg into an unspecified orbit.[49]

A less successful rapid-response launch technology, the railgun, employs electromagnetic propulsion. One such device was apparently developed at University of Texas's Centre for Electromechanics. Another was built at the US Army Research and Development Centre. The US Navy initiated a further electromagnetic railgun program, beginning in 2005, the design intended in its mature state to have a 370 km range.[50] Applying the broad *½ R (range)* rule for converting horizontal capacity into vertical range, this device when fully developed could accelerate a projectile of undisclosed mass to approximately 185 km vertically. That launch capacity could be adequate to launch a microsatellite, subject to the constraints identified below, although its orbit would be below 185 km because of the requirement to launch at an angle sufficient to impart escape velocity and permit entry into a low-Earth orbit.

Regardless of developments designed to increase the range and efficiency of railgun devices, the compression and related forces exerted by rapid acceleration would probably limit material thus accelerated to, the most robust of substances. One estimate of railgun compression was that it would be equal to 1 000 g.[51] Even if that estimate was overstated by an order of magnitude, a force of 100 g would present design challenges for any payload containing sensitive electronics and other delicate components.

Railgun technologies other than those enabled by electromagnetic energy have been proposed and might be developed beyond the modelling stage but the compression-acceleration challenge will persist unless the gun-launch can be coupled with other forms of on-board propulsion.

Railguns and similar technologies cannot be totally ignored in the context of space warfare because they might be converted and used for purposes other than spacecraft launch. On 9 February 2012 the US corporation Raytheon announced it had been awarded a contract by US Navy Sea Systems Command for preliminary design of a '*pulsed power system*'. The associated railgun projectile was predicted to travel up to ~350 km in less than six minutes and to exit the dense atmosphere before hitting its target at a speed of ~ 5 500 km/h.[52] This might imply a capacity to acquire and attack an *approaching* spacecraft, including a missile. However, orbiting spacecraft in the 200 – 500 km altitude band will travel at ~ 27 000 - 28 000 km/h, and so any such object fired after the spacecraft had passed overhead would be too slow to catch-up with it.

Ballistic spacecraft with Earth-centric orbits encounter several unavoidable phenomena which, singly and in combination, change and degrade their orbits. To keep spacecraft approximately in their planned orbits, satellite operators negate those effects normally by activating boosters and thrusters attached to each satellite. This opens the possibility of booster and thruster technologies being further developed to permit satellite manoeuvres, allowing a satellite to evade an aggressor or, if the satellite is itself the aggressor, to rapidly approach an intended target. Evolving technologies indicate that those capabilities will become increasingly viable as, '*... these devices provide the starting point for future developments to achieve much higher power levels.*'[53]

In August 2011 a published description of technical problems encountered by a malfunctioning communications satellite, the US Advanced Extremely High Frequency Satellite, included the statement, '*The satellite's main engine was supposed to produce 100 pounds of thrust* [~445 N] *while burning hydrazine fuel and nitrogen tetroxide ...* '[54] This indicated a potent manoeuvring capacity under some circumstances and it may have been convertible to function as one component of a suite of defensive capabilities.

On 12 January 2012 the US corporation Boeing announced it was developing a '*mission concept study*' to evaluate concepts combining high-power solar arrays with advanced electric thrusters to power spacecraft and payloads toward high Earth orbits and deep space destinations. The study was being undertaken by Boeing's Phantom Works with support from Boeing's Space and Intelligence Systems. The announcement preceded advice that a similar study was being advanced by Northrop Grumman Aerospace Systems, with the proposed propulsion system to be scalable to 300 kW [300 000 Nm/s].[55] The French company Snecma has reported on its design of a 20 kW [20 000 Nm/s] satellite thruster.[56]

These technologies are natural extensions of established satellite station-keeping principles. The basic knowledge and engineering competencies are therefore widespread within the international community. Qualitative variations will likely exist and will favour the more technologically advanced States.

Satellite thruster technology is likely to become a two-edged sword. It could offer some capacity for space vehicles to avoid and confuse outer space weapons as well as to avoid foreseen items of space debris. Simultaneously, it provides opportunities for space vehicles to acquire a new role as space aggressors and could also be adapted to make advanced missiles more difficult to attack.

Contemporary literature almost invariably characterises space debris as a hazard to be controlled or mitigated. That was not always the case. As already recorded, on 16 October 1957, 12 days after Sputnik 1's launch, the US Air Force fired a cloud of pellets to an attitude of 87 km. In May 1963 the US conducted Project West Ford, an experiment intended to create an orbital ring of filaments from which radio transmissions could be reflected. This involved launching into orbit and dispersing ~480 000 000 copper filaments at an altitude of 3 170 km. Additionally, in the mid 1980s the US Office of Technology Assessment speculated it '*should be possible*' to build a directional fragmentation warhead, similar to a Claymore mine, for launch into outer space where it would project 10^5 x 1 g pellets in a pattern covering a 100 x 100 m area. Space-based guns and rockets, their projectiles impliedly based on kinetic principles, were also contemplated.[57] Other States conceivably had similar ambitions and possibly programs.

The less discriminatory applications of those technologies are no longer practicable. But, for an aggressor there remains the possibility that selective and precisely applied damage to a satellite might fortuitously be attributed to collisions with space debris, subject to limitations already mentioned. At least one functioning satellite (*Iridium 33*, February 2009) has already been destroyed in a debris-related collision, although in that case the debris had been identified and was being tracked, albeit not accurately enough to prevent a collision. Many other, and possibly most, spacecraft have been struck by orbiting debris but were neither destroyed nor damaged to the point where their functional capacity observably deteriorated.

Relatively low-powered lasers, either terrestrially or space based, can be deployed against the imaging sensors carried on some categories of satellites. Those sensors can be either dazzled (affected temporarily) or blinded (affected permanently). There is little or no doubt almost every State already has or could rapidly acquire this capacity.

Successful applications of the low-powered laser technology have been alleged. The US has claimed the USSR blinded one US satellite in October 1975 and another in November of the same year.[58] In late 2006 US National Reconnaissance Office Director,

Donald Kerr, claimed that the PRC had recently used a laser to '*illuminate*' a US satellite.

The strategic or tactical potential to dazzle or blind spacecraft sensors can nevertheless be overstated. To dazzle (or blind) a satellite sensor, the laser must be positioned so that its transmitted energy contacts the sensor's lens either directly or through optical and related pathways integral to the sensor. Therefore a dazzling laser in outer space would normally have to be in an orbit lower than that of the Earth-observing sensor it was attempting to affect. There is a relationship between orbital altitude and velocity (the lower the orbit the greater the velocity) and so opportunities for successful encounters by orbiting lasers seeking high-value targets would be time-constrained unless the laser was mounted on an agile space vehicle, one able to absorb resultant fuel consumption penalties and continue operating or, alternatively, on a spacecraft designed to be sacrificed after a very limited number of attacks.

It can also be difficult to aim a laser so its energy reaches any given satellite's sensor. Most sensors designed to collect data for conversion into images are either mechanically or electronically gimballed. The sensors thus move in patterns known as crosstrack, whiskbroom and pushbroom, as they collect data. Consequently, the area within view of a sensor at any given time, its field-of-view and swath, changes constantly as it senses within a wider area commonly referred to as its field-of-regard. It follows that the initiator of a blinding or dazzling laser attack against a space-based sensor would be significantly disadvantaged without prior and intricate knowledge of the operation of a target sensor, noting that the sensor's operation might be periodically changed by reprogramming.

Additionally, imaging sensors are commonly equipped with electrical or mechanical shutter mechanisms, so the dazzling laser must engage the sensor when a shutter is open. (This form of sensor control is distinct from the diplomatic, financial and legal initiatives taken by some States to limit the distribution of imagery products, a practice also referred to as 'shutter control'.) In addition, sensor designers have been developing defences against laser light, such as inclusion of liquid crystal protection or, more recently, graphene dispersion technology.

To reach a sensor carried on a spacecraft, terrestrially based lasers will be limited to wavelengths able to penetrate the dense terrestrial atmosphere, although some flexibility could be obtained by placement of such instruments on areas of high terrain and low humidity or on aircraft.

Taking all of these constraints into account, it seems probable the most effective application for laser dazzlers or blinders into the indefinite future will be as terrestrial point-defence mechanisms. Employed in the point-defence role they will have some utility in limiting the imaging from space of highly sensitive fixed installations or tactically valuable locations, such as the construction of a ship or installation, or the concentration of troops and supplies in preparation for an attack.

Evolving Defences Against Space Weapons

Spacecraft can avoid, resist or reduce damage caused by energy directed toward them. They can be designed with external profiles based on curved surfaces instead of flat or plane surfaces, resulting in the distribution and therefore the dilution of incoming destructive energy across a larger area. Spin-stabilised satellites have a particular advantage in this regard. Sheathing materials with insulating properties have been developed for spacecraft. This has increased the magnitude of challenges already associated with the design and application of weapons intended to cause melting or ablation in outer space.

Satellites and other spacecraft can also be coated with materials known to absorb laser-induced energy. Graphite was one early coating material with absorbent properties, but others are likely to have been discovered and tested both for outer space and terrestrial applications. Reflective coatings may offer additional and effective partial protection against laser attack. (However, panels of photovoltaic cells might not be compatible with most of these coatings.) New forms of electronic 'invisibility' cloaking are being developed and these may have some applications in camouflaging spacecraft.

Almost every spacecraft is equipped with a capacity to affect minor station-keeping and orbital correction manoeuvres. The technologies involved in enabling these manoeuvres can be programmed to move a spacecraft in response to detected threats. The potential for spacecraft to utilise high-powered thrusters and related technologies exists and is increasing, as I have mentioned.

Electronic components designed for use in satellites are typically radiation hardened to survive predictable natural hazards. High-value satellites of strategic significance are likely to be radiation hardened beyond the levels anticipated from natural hazards. Consequently, attacks employing radiation to achieve a disabling effect may have limited capacity to harm the most significant military satellites and have potentially maximum adverse consequences only for some commercial satellites.

In the case of devices intended to jam or otherwise interfere with data transmission to and from spacecraft, one development favouring satellites is that radio transmissions are seldom omnidirectional; instead, data are commonly sent and received within tightly confined beams which are steerable. Consequently, to achieve radio frequency interference, the jamming transmitter might have to either interact with one of the satellite's beams or to prevent its transponders and related antennae from receiving or transmitting data. Optical protection devices were mentioned previously.

A further factor potentially limiting the utility of most categories of space weapon, particularly when deployed against ballistic missiles, is that an intended target may have been armed with a salvage fuse designed to detonate and create a debris cloud if the spacecraft is attacked or endangered. This potentially limits the advantage otherwise conferred by allegedly non-fragmenting attacks.

In addition to all other factors limiting the effectiveness of attacks against spacecraft is that the likelihood of undetected aggression in outer space is becoming increasingly remote as the technological capacity and associated commitment to observation of space assets is increasing. Detection and identification of aggression in outer space is likely to result in retaliation if a victim State or its strategic allies are capable of responding.

Consequences

Notwithstanding spacecraft protective technologies and strategies, eight categories of weapon are actually or potentially able to harm an orbiting space vehicle. They are: nuclear weapons including NDRE, kinetic devices, co-orbital satellites, microwave weapons, other radiofrequency weapons, space planes (as primary enablers), devices enabled by hypervelocity vehicles, and relatively low-powered weapons such as dazzling lasers.

The amount of energy required for any weapon to damage a space vehicle will vary primarily in accordance with the physical distance between the weapon and its target, the timescale in which termination of the target spacecraft's capacity to function is required, the amount of damage or harm needed to stop or inhibit the vehicle's ability to perform its intended mission and the capacity of a target to resist an attack.

Nuclear bombs are the most reliable space weapon available in terms of the probability of success in destroying a designated target. They do not require precise spatial positioning to be effective whereas all practicable alternatives are dependent on more advanced and complicated positioning and guidance systems. However, nuclear weapons are likely to cause widespread and enduring collateral damage, from both debris and persistent NDRE phenomena.

Missiles carrying kinetic and related weapons have a record of success against space vehicles. The technology, already convincingly demonstrated by Russia, the PRC, Japan and the US, is likely to proliferate. Several additional States, notably India and Israel, appear to be consciously pursuing the development or acquisition of anti-satellite kinetic missiles. There are no obvious impediments to this or closely related technologies being developed to function as far as the GEO orbits or beyond.

Co-orbiting vehicles will be among the most versatile of space weapons given they can potentially host a range of aggressive technologies allowing them to attack and debilitate a space vehicle. Additionally, some co-orbital vehicles will have the capacity to forcibly deorbit spacecraft except for those spacecraft equipped with thrusters more potent than those being carried by the attacking vehicle.

Two thus far immature developments could enhance the effectiveness of co-orbital weapons in outer space. Space planes could, and seem likely to, evolve into vehicles allowing significant flexibility for deployment in outer space. They can be expected to enable or enhance most or all functions now assigned to smaller co-orbital vehicles.

It is highly likely weapons associated with burning, ablating and radiation in outer space will become practicable only if power is available from collocated nuclear reactors. The radiation characteristics of both high-powered weapons and of nuclear reactors in outer space would provide readily discernible signatures and likely give positioning data to all potential opponents. The capacity to enable such space weapons using alternative energy sources, such as solar and chemical fuel cells, is marginal at best and likely to remain so.

By 2030 some terrestrially based laser technologies could conceivably be developed for use as a high-powered weapons against spacecraft. However, a more confident assessment is that terrestrial lasers will be used for blinding and dazzling roles under specific conditions, such as in point-defence roles.

The deployment and utilisation of most weapons against spacecraft is an imprecise art. Apart from attacks using nuclear bombs some forms of missiles and possibly co-orbiting vehicles, successful attacks against spacecraft cannot be reliably forecast. The potential for retaliation for an attack against a spacecraft will remain high and the retaliation will not necessarily be initiated in outer space.

No State appears to have a monopoly on, and might not have even a significant advantage through, access to demonstrably effective weapons available to be used against spacecraft. The concept of any single State capturing and dominating near-Earth space is therefore probably unsound. However, several States have capacity to render near-Earth space unusable by intensifying the density of orbiting debris. If that happened the space region would be denied to the aggressor also. It would be the classic example of a nation State shooting itself in the foot.

There is a possibility that highly significant new developments likely to impact warfare in outer space are occurring in the security classified environment, but this is unlikely. First, such achievements would need to occur in the face of obstacles so daunting as to have defied resolution for decades. No information found pointed to the resolution of whole-of-system obstacles, some of which stem from limitations imposed by basic laws of physics. A common tendency with newly-developed weapons, particularly in the US, Europe and India, has been to announce and at times overstate developmental achievements. Only sensitive operational specifications have been routinely concealed. I located no such claims regarding major and unanticipated developments of space weapons.

Indications have emerged that the US, for example, is beginning to accept this situation. The US Department of Defense (2010) recorded;

> *... the Department of Defense will realign spending away from defenses planned to rely on currently immature technology, away from technologies that require unrealistic concepts of operations in order to be effective, and away from technologies intended to defeat adversarial missile threats that do not exist and are not expected to evolve in the near to midterm. These considerations led to decisions to terminate both the Multiple Kill Vehicle and Kinetic Interceptor programs and to shift the Airborne Laser to a technical demonstrator program in the FY 2010 budget.*[59]

The implications of international law for space warfare and space weapons are now reviewed in Chapter 6, *International Law and Outer Space.* Despite significant efforts invested by jurists and diplomats to develop a legal regime relating to the conduct of affairs in outer space, the discussion following discloses the fragility of international law in this geopolitical setting.

During the Second World War Germany unveiled the Vergeltungswaffe-2 *(V2) missile. It could reach altitudes higher than 80 km and so became the first anthropogenic item to enter outer space as defined in this work.*

Photo courtesy, imgkid.com/V2-rocket-ww2.shtml

There is minimal distinction between a space launch vehicle and a long-range missile. A space launch vehicle with a payload of more than 1 000 kg could be modified to take a powerful nuclear weapon into orbit. The vehicle shown is a European Ariane 5 with a payload capacity (into low Earth orbit) of ~ 20 tonnes (There is no implication that this launch vehicle was designed to be used for other than peaceful purposes, has ever been used for other than peaceful purposes or that plans exist to modify it for other than peaceful purposes.)'

Photo iStockphoto (22476768)

Satellites are orbiting with many other spacecraft supporting navigation, precision timing, communications, intelligence collection, agriculture, military activities, mapping, search-and-rescue and other functions. Even a common meteorological satellite, can provide data on weather, climate, soil salinity, soil moisture, forest fires, ice storms, cyclones, floods and volcanic ash dispersal. The consequences of loss or degradation to all space services would be globally catastrophic.
Photo iStockphoto 907089

Advanced space technologies are not confined to any nation State. The PRC space station and unmanned space craft, shown here, appear benign and can be fairly represented as a scientific initiative. But they also represent technical ability to approach an orbiting satellite with the objective of harming it or ascertaining its potential as a future target. Co-orbital technologies have been under development for decades.
Photo Getty Images 166609668

Chapter 6. International Law and Outer Space

In this chapter I primarily address the way international law operates to discourage or contain warfare, particularly warfare in outer space. Discussion following includes references to legal terms, such as *ratify* or *accede*, which have specific meanings and applications within the framework of international law as it applies to treaties and similar arrangements. For the convenience of readers without legal training the terminology is summarised in endnote 1.[1]

Acknowledging the challenges and controversies associated with any attempt to comprehensively define, let alone identify, international law the following discussion is focussed upon international treaties and on resolutions adopted by the United Nations (UN) General Assembly because they are readily identifiable and verifiable. I have not considered international law relating to dealings between private individuals or to minor commercial dealings, sometimes referred to as *private* international law.

During centuries past, spasmodic attempts were made to alleviate the worst consequences of warfare, examples being the previously referenced Second Lateran Council's ban in 1139 CE on '*incendiarism*', or the *Strasbourg Agreement* of 1675 which banned poisonous bullets in combat between France and Germany. However, by the early 19th century jurists had effectively defined war as a '*legal condition in which it was lawful for the two contending states to rain death and destruction upon one another.*'[2]

A predecessor to a more modern concept of a law of warfare emerged with the *Declaration Respecting Maritime Law,* (Maritime Declaration (1856)) banning privateering and addressing other maritime issues. It was followed by the *Convention for the Amelioration of the condition of the Wounded in Armies in the Field,* (Geneva Convention (1864)). The concept of avoiding warfare *per se* was seldom raised.

Avoidance of war was eventually addressed by the *Convention for the Pacific Settlement of International Disputes, 1900* (Hague I). It required, Article 1, '*Signatory powers agree to use their best efforts to ensure the pacific settlement of international differences.*' It continued with Articles 2 - 8 on mediation and Articles 15 – 19 on arbitration. Those concepts failed to prevent the outbreak of the First World War in 1914. However, following that conflict, the theme of avoiding warfare was resuscitated and pursued through formation of the League of Nations.

The League of Nations

The 1919 *Covenant of the League of Nations* (the Covenant) was in some respects a template for the later *Charter of the United Nations* (the UN Charter). The Covenant's prologue emphasised the need for international peace and security based on regard for justice and international law. It structured the League of Nations as an assembly with a council and a secretariat. In arrangements comparable to those of the subsequent UN Security Council, the League of Nations Council was to include permanently the representatives of principal allied and associated powers (identified within the context of the First World War) supplemented by members selected through its Assembly. The Covenant also included arrangements for the League of Nations to oversee specialised international bureaux, as the UN did subsequently. However, within the League of Nations, votes endorsing matters of substance within both the Assembly and the Council, had to be unanimous.

The League required armaments of all nations be reduced to the lowest point consistent with national safety and the enforcement of international obligations. This was not a mere idealistic ambition; specific implementation proposals and arrangements to monitor its progress were activated. Members also agreed that if they resorted to war against another member in disregard of those obligations they would be deemed to have committed an act of war against all members of the League.

The League of Nations contributed measurably to global stability. It defused potential conflicts between Sweden and Finland (1920) and between Greece and Bulgaria (1925). The League arranged a financial package preventing Austria from defaulting on debts. It conceivably created the environment in which the US and 14 other nations agreed in August 1928 to renounce '*recourse to war for the solution of international controversies*', through the *Peace Pact of Paris* (Kellogg-Briand Pact). The Pact was eventually endorsed by 62 nations with signatories including Australia, Italy and Japan.

The League of Nations was unable to intervene usefully in events such as the Japanese invasion of Manchuria in 1931 or the rearmament of Germany during the 1930s. It also failed to prevent illegal military-related initiatives by several States, although to some extent these actions were undertaken clandestinely, reflecting adversely upon the States involved rather than the League.[3] The League of Nations could not prevent Stalinist purges and domestic terror campaigns in the USSR.

The League suspended activity during the Second World War but did not formally disband until 12 April 1946 having not during its existence successfully intervened in any military conflict involving a powerful State.

The United Nations

International laws formulated to discourage, curtail or regulate warlike behaviours are now substantially contained in the *Charter of the United Nations* (UN Charter, 1945) and the associated *Statute of the International Court of Justice* together with the Geneva Conventions and their protocols. However, the Geneva Conventions of 1949 maintained a focus on international humanitarian law and were not primarily intended to prevent or mitigate conflict itself as was the UN Charter.[4]

The very existence of this Charter offered an impression of progress having been made with the containment of international violence. But a more realistic evaluation might be that, '*In a melancholy repetition of the experience of the League of Nations, the UN very quickly foundered on the hard realities of global politics.*'[5]

From the outset it was clear that the UN wanted to avoid warfare. The opening sentence to the preamble of the UN Charter read;

> '*We the peoples of the United Nations determined ... to save succeeding generations from the scourges of war...*'

Article 1 of the UN Charter, recorded its purpose as;

> '*To maintain international peace and security, and to that end: to take effective collective measures for the prevention and removal of threats to the peace, and for the suppression of acts of aggression or other breaches of the peace, and to bring about by peaceful means, and in conformity with the principles of justice and international law, adjustment or settlement of international disputes or situations which might lead to a breach of the peace; ...*'

Components of Article 2 of the UN Charter were yet more direct.

a. Paragraph 3. *All Members shall settle their international disputes by peaceful means in such a manner that international peace and security, and justice, are not endangered.*

b. Paragraph 4. *All members shall refrain in their international relations from the threat or use of force against the territorial integrity or political independence of any state, or in any manner inconsistent with the Purposes of the United Nations.*
c. Paragraph 6. *The Organization shall ensure that states which are not members of the United Nations act in accordance with these Principles as far as may be necessary for the maintenance of international peace and security.'*

In addition, the Security Council declared forcible reprisals to be a violation of the UN Charter.[6] The General Assembly endorsed that position.[7]

The UN Charter and supplementary resolutions thus closely approached the point at which it might have been possible to categorically declare warfare by or between its member States to be illegal, but the unambiguous language to support this step was not taken.

The UN Charter made the Security Council the preeminent forum in most matters concerning peace and security. Member States confer on it primary responsibility for the maintenance of international peace and security and they agree to accept and carry out its decisions. (Article 24) Additionally, (Article 12) while the Security Council is exercising its authority in respect of any dispute or situation the General Assembly may not make any recommendations with regard to that dispute or situation unless the Security Council requests.

Article 51 of the UN Charter appears to offer members of the United Nations some capacity for independent action in self defence by claiming, '*Nothing in the present Charter shall impair the inherent right of individual or collective self defence if an armed attack occurs against a member of the United Nations.*'

But Article 51 continues by imposing two conditions:

a. The right of individual or collective self defence exists only until the Security Council has taken measures necessary to maintain international peace and security, and
b. measures taken by members in the exercise of this right of individual or collective self defence are to be immediately reported to the Security Council and are not to affect its authority and responsibilities.

The UN Charter permits the existence of regional arrangements or agencies for dealing with matters relating to the maintenance of international peace and security, provided they are consistent with the purposes and principles of the United Nations. However, Article 53, paragraph 1, also provides that the Security Council shall, where appropriate, '*utilise such regional arrangements or agencies for enforcement action under its authority.*' Other components of the UN Charter confer either additional powers upon the Security Council or create obligations toward it by members of the General Assembly.

The Security Council comprises 15 members, ten being elected by the General Assembly for a two-year term in office, with five (the PRC, France, Russia, Britain and the US) being permanent members. Decisions of the Security Council on matters, other than procedural matters, require the concurring votes of the five permanent members, provided a party to a dispute must abstain from voting under defined circumstances.

Ironically, the pervasive influence of the Security Council, combined with the challenge of persuading all five permanent members to agree on many issues, may be leading to limitations on its power. There is evidence of at least one alternative arrangement having emerged.

On 23 September 2008 the UN Secretary-General, Ban Ki-Moon, signed the *UN-NATO Cooperative Declaration* with North Atlantic Treaty Organisation (NATO) Secretary-General Jaap de Hoop-Scheffler. I could find no record of the matter having been referred to the Security Council, and Russia was reported as considering the arrangement 'illegal'. The content of the document apparently was and remains either security classified or of very limited distribution.[8]

The UN Charter was framed to make it difficult or impossible to curtail powers granted to the permanent members of the Security Council, or to amend permanent Security Council membership, other possibly than by adding to the number of permanent Security Council members.[9] However, the difficult or impossible did occur once when a permanent member of the Security Council was expelled and replaced by another member.

Following a campaign by members of the General Assembly, in 1971 a vote resulted in the PRC replacing Taiwan (the Republic of China) as a member of the United Nations and as a permanent member of the Security Council.[10]

Since the 1970s, the influence of some other States, not permanent members of the Security Council, has been expanding. Those States include India, Japan and Germany. Meanwhile, Britain has retained its role as a permanent member of the Security Council despite now having the seventh to ninth largest economy in the world (its ranking depending on the measurement methodology employed) and being ranked in that context behind Japan, Germany and Brazil, although economic size is not the only arbiter of geopolitical influence.

The UN has undertaken various peace-related activities since 1945. At February 2014 it had a total of 16 peacekeeping forces deployed. The UN has also attempted to constrain the proliferation of weapons of mass construction, and has been engaged in movements opposing the use of some conventional weapons.

The UN has additionally authorised the use of armed force on several occasions. An intervention in the Korean War (1950-53) was successful in that the independence of the RoK was preserved. Armed intervention in the Republic of Haiti was endorsed by the Security Council in 1994. Military intervention was effectively authorised (*'use all necessary means'*) against Iraq in 1990.

However, also since 1945 a variety of justifications has been used by United Nations member States, including permanent members of the Security Council, to justify their own military actions when not specifically endorsed by the United Nations. France and Britain portrayed participation in the Suez crisis of 1956 as a peacekeeping operation. The US claimed defence of US nationals located internationally as part of its justification for interventions in the Republic of Dominica (1965), Grenada (1983) and the Republic of Panama (1989 – 90).

The UN did not intervene effectively in the US and its allies' war in Vietnam (c.1964 – 1975); the USSR's invasion of Afghanistan (1979 – 1989); or the PRC's mass killing of its own citizens during the period beginning in 1949. The UN response to Iraq's aggression toward Iran and Kuwait was slow, although military intervention was authorised following the Iraqi invasion of Kuwait in 1990. UN interventions were either late or failed to prevent deaths in the Republic of Burundi, the former Yugoslavia, the Kingdom of Cambodia, and in the Lebanese Republic.

Since 1945 there has not been a conflict to compare with the First or Second World Wars, if measured by the number of fatal casualties. However, as stockpiles of nuclear weapons grew, the Cold War (c. 1945 - 1991) had potential to cause more casualties than the First and Second World Wars combined. It might never be possible to gain general agreement on the relative influence of many factors associated with ending the Cold War; however, I found no evidence to suggest the UN played a pivotal role in bringing it to a close.

In summary, the UN does not appear to have been significantly more effective in inhibiting the military and related activities of powerful States than was the League of Nations, an organisation commonly regarded as having failed to achieve its purpose.

Law and Warfare in Outer Space

The UN can foster international arrangements regulating behaviour in regions which no State can legitimately claim to own. The UN has thus been involved with consolidating centuries-old traditions of activity on the high seas and also in the regulation of airspace above the high seas, both environments being global property according to international law. Consequently, the notion of international law embracing acts or omissions in outer space has not been controversial, although aspects of the law applied to outer space have been and remain subject to debate.

International interest in and support for the development of space law may have increased over the past six decades. Between December 1958 and December 2011, the UN General Assembly adopted a total of 111 space-related resolutions. During each decade, beginning with 1958, the number of such resolutions, with one minor exception, increased.[11] Over a similar period, membership of the UN also increased, from 82 members in 1958 to 193 in January 2014. Accordingly, there has been an approximate relationship between the expanding number of UN member States and the number of resolutions concerning outer space. However, the number of States ratifying or acceding to the various treaty-level arrangements has, paradoxically, been declining as illustrated by Table 6 – 1. *Ratifications and Accessions to Outer Space Treaties,* below.

Table 6 – 1. Ratifications and Accessions to Outer Space Treaties (Compiled January 2014)[12]

Treaty-level Arrangement	Year of Entry into Force	Number of Ratifications or Accessions (Including Security Council Members)	Number of Permanent Members of Security Council Ratifying or Acceding
Outer Space Treaty	1967	102	5
Rescue Agreement	1968	92	5
Liability Convention	1972	89	5
Registration Convention	1976	60	5
Moon Agreement	1984	15	0

A need to regulate activities in outer space had been recognised even before the negotiations had started for any outer-space-related treaty. In August 1957 the US proposed the development of an inspection system for outer space as one component of a proposal for partial disarmament, but this was rejected by the USSR. Between 1960 and 1962 several USSR proposals for disarmament, including provisions relating to outer space, were rejected by the US. On 17 October 1963 the UN General Assembly adopted Resolution

1884, calling on all States to refrain from introducing weapons of mass destruction into outer space.

On 16 June 1966 both the US and USSR introduced draft treaties addressing broadly-based space-related issues. The US version dealt only with celestial bodies and the USSR's draft included all of outer space. Several related issues were resolved through consultations between representatives of the US and USSR. This led to a resolution (General Assembly Resolution 2222 (XXI) of 19 December 1966) commending the *Treaty on Principles Governing the Activities of States in the Exploration and Use of Outer Space, including the Moon and Other Celestial Bodies,* (Outer Space Treaty (1967)).

The Outer Space Treaty was the first of its kind. When conceived and drafted during the Cold War the US and USSR were so-called superpowers. The treaty document which came into force on 10 October 1967 consequently reflected both the idealism of a nascent space age as well as the geopolitical reality of two powerful, ambitious and opposing States, each capable of exerting global influence.

Complicated and incompatible objectives of the US and USSR, affecting the Treaty, included a desire to permit the use of intercontinental missiles should a tenuous Cold War stalemate collapse, to endorse the use of satellites for collecting intelligence from each other or from any other State, and to establish constraints on unpredictable activities beyond Earth, including specifically upon the Moon. The Outer Space Treaty also included elements designed to endorse the cooperative and peaceful utilisation of outer space. Underlying this farrago of aims was an implied determination to avoid defining the altitude at which outer space began, thus allowing some activities to be categorised as having occurred either in outer space or in the terrestrial environment as it suited the circumstances of any given incident.

Despite compromises and limitations, the Outer Space Treaty has remained at the epicentre of international law affecting outer space. It provided the foundation upon which subsequent international laws were constructed. And it seems almost certain to retain that status for the indefinite future. No such agreement applying to outer space is likely to be negotiated in the early decades of the twenty-first century because there has been a move away from treaties toward voluntary and non-binding arrangements.

The Treaty's prologue was idealistic and included claims to have been; '*Inspired by the great prospects opening up before mankind ... , Recognising the common interests of all mankind in the progress of the exploration and use of outer space for peaceful purposes ... , Believing that the exploration and use of outer space should be carried on for the benefit of all peoples ... ,* [and] *Desiring to contribute to broad international cooperation ...*'

This theme was reflected in its text which required(Article I); the '*exploration and use of outer space, including the Moon and other celestial bodies, shall be carried out for the benefit and in the interests of all countries...* [and there shall be]... *freedom of scientific investigation in outer space.*' Article II stated that outer space, including the Moon, '*is not subject to national appropriation by claims of sovereignty by means of use or occupation, or by any other means*.' Article III required that activities in outer space remain in accordance with international law, including the Charter of the UN.

Article IV of the Treaty was more selective and specific. It:

a. Prohibited placing in orbit around the Earth any objects carrying nuclear weapons or any other kinds of weapons of mass destruction, or installing such weapons on celestial bodies, or stationing such weapons in outer space in any other manner.
b. Required the Moon and other celestial bodies be used by all States Parties to the Treaty exclusively for peaceful purposes. The establishment of military bases, installations and fortifications, the testing of any types of weapons and the conduct of military manoeuvres on celestial bodies was '*forbidden*'.

States Parties to the Treaty were made responsible for carrying out activities in outer space in accordance with the treaty and that responsibility was extended to include non-government entities or agencies of each State Party.

The Treaty. Article VII, also made each State Party '*that launches or procures the launching of an object into outer space, including the moon and other celestial bodies, and each State Party from whose territory or facility an object is launched, is internationally liable for damage to another State Party to the Treaty or to its natural or judicial persons by such object or its component parts on the Earth, in air space or in outer space, including the moon and other celestial bodies.*'

The Outer Space Treaty, if adhered to, therefore provided a legal framework for regulating human activities in space. It also supported each of the outer space-related treaties that followed. However, its defects and omissions also influenced and were reflected in subsequent international law as it evolved.

As I have stated, the Outer Space Treaty did not define the geographic boundary between terrestrial environments and the outer space region. Consequently, doubt has persisted over whether some activities occur in outer space or in the terrestrial atmosphere. It did not establish an outer space regulatory body, comparable for example with the International Civil Aviation Organisation (ICAO), and therefore made no provision for adjustments to take account of revised circumstances or new technologies. (A United Nations committee, the Committee on the Peaceful Uses of Outer Space (COPUOS) fills this function in part.) It did not provide for any elaborating or subordinate regulations and left States parties to resolve between themselves any issues arising.

While the Treaty banned the placing in orbit around Earth any objects carrying nuclear weapons and weapons of mass destruction it remained silent regarding all other forms of weapon and did not address the potential for terrestrial weapons to be aimed at orbiting spacecraft. Additionally, 'orbit' was not defined thus the Treaty could be interpreted as allowing intercontinental missiles to travel travel above the altitudes of some orbiting satellites, across oceans and beyond, without completing an orbit. The prohibition against nuclear weapons did not appear to ban the orbiting of nuclear reactors intended to provide enabling power for other categories of weapons. Consequently, the potential for aggressive competitiveness to include the outer space region was retained.

The subsequent *Convention on International Liability for Damage Caused by Space Objects* (Liability Convention (1972)) was another jurisprudential product of the Cold War. By 1972 both the US and USSR had developed kinetic and nuclear anti-satellite weapons. The US had launched one nuclear reactor into outer space and the USSR had launched several. Humans were already being transported to and through outer space, with the first spacewalk, by AI Leonov, occurring on 18 March 1965 and the first moonwalk, by N Armstrong, on 20 July 1969. A transatlantic telecast had been relayed by US satellite *Telstar 1* on 10 July 1962. The relevance of outer space to strategic military functions had not diminished but a range of space-dependent societal and scientific functions was emerging.

The prologue to the Liability Convention was in part idealistic. It included recognition of '*the common interest of all mankind ...* ', but it continued in a more pragmatic tone to recognise, '*...the need to elaborate effective international rules and procedures concerning liability for damage caused by space objects and to ensure, in particular, the prompt payment under the terms of this Convention of a full and equitable measure of compensation to victims of such damage.*' The text imposed obligations upon launching States and others, including third parties, and described processes for apportionment of damage caused.

However, the circumstances in which liability could vest upon any party were caveated. Article III stated;

> *'In the event of damage being caused elsewhere than on the surface of the earth to a space object of one launching State or to persons or property on board such a space object by a space object of another launching State, the latter shall be liable only if the damage is due to its fault or the fault of persons for whom it is responsible.'*

Article VI, imposed unconditional liability for damage only when a State had not been operating in accordance with international law including with the UN Charter.

However, the Convention focused on damage caused by '*space objects*'. A space object, contextually, is a discernible, physical object (' *component parts of a space object as well as its launch vehicle and parts thereof*' (Article 1)). In effect, the Convention primarily envisaged physical collisions in outer space. This interpretation could include damage caused by kinetic weapons, some co-orbital weapons and similar devices. However, liability for damage to space objects caused by directed-energy or radiofrequency weapons deployed in space was not apparently foreseen. The potential for terrestrially-based weapons to project beams into outer space also escaped notice or was intentionally ignored when the Convention was drafted.

An additional complication involves the pursuit of claims for compensation. Of the total 28 articles in the Convention, 16 are dedicated to compensatory processes. The various avenues provide for choice but also for obfuscation. By 1993 one researcher was able to identify 57 international instruments dealing in some way with the settlement of disputes regarding space activities.[13]

So far as I could determine, by early 2014 compensation had not been paid for the destruction of an active satellite, *Iridium 33,* following collision with an inoperative satellite, *Cosmos 2251,* on 11 February 2009. The Iridium consortium claimed compensation but Russia argued that *Cosmos 2251* had not been under control and *Iridium 33* should have been manoeuvred safely away.[14] The incident emphasised the fragility of the core compensation provisions in this Convention.

The prologue to the later *Convention on Registration of Objects Launched into Outer Space,* (the Registration Convention (1975)) recognised;

> *'the common interest of all mankind ...'* before focussing on the need to, *'provide for States Parties additional means and procedures to assist in the identification of space objects,'* as well as believing that, *'a mandatory system of registering objects launched into outer space would, in particular, assist in their identification and would contribute to the application and development of international law ... '*

A central provision of the Registration Convention was its Article II;

> *'When a space object is launched into Earth orbit or beyond, the launching State shall register the space object by means of an entry in an appropriate registry which it shall maintain. Each launching State shall inform the Secretary-General of the United Nations of the establishment of such a registry.'*

These requirements, when complied with, provide a mechanism for understanding and observing the orbits of spacecraft. Such data could limit the potential for collisions between spacecraft or reduce international tensions associated with close approaches by spacecraft of one State toward those of another State, often referred to as *conjunctions*. The same data could also provide States with information to reduce the possibility of confusing a benign spacecraft with a missile or other weapon.

Conversely, the publication of all orbital specifications for every space vehicle would reveal the existence of deliberately unobtrusive national assets in outer space, making them more vulnerable to interference in periods of tension. This potential disadvantage probably created temptations to not register spacecraft if they could be camouflaged or otherwise made difficult to detect. Evidence points to the temptation having been translated into action.

Viikari (2008) has claimed:

> *'The practice of states (even the states parties to the Registration Convention) in registering space objects differs in many respects. For instance, the time for submission of information to the UN varies from weeks after launch to years (the average being two to three months). As concerns the particular information which is registered, the only coherent data provided seems to be the name of the launching state(s). There is no mechanism for controlling the accuracy of the information provided. A telling fact is that apparently no state has ever registered a launch of a space object as having a military purpose. About one-third of space objects launched are not registered with the UN at all. Furthermore, only about 50 per cent of re-entries of space objects (of States parties to the Registration Convention) to the Earth's atmosphere are notified to the UN.'*[15]

It was not clear how Viikari quantified unregistered launches, or how her related quantifications might be verified. However, other sources support the broad assertion that some spacecraft are unregistered. The French space agency, Centre National d'Etudes Spaciales (CNES), in 2011, reported some nine per cent of satellites in orbit had not been registered.[16] The UN Office for Outer Space Affairs has estimated that, *'approximately 93.5%' of all functional space objects have been registered with the Secretary General.*[17] The statistical precision in this last example implied knowledge of which space objects had not been registered. Another observer, also impliedly with access to detailed data, claimed that by 2011 out of 3 360 launches USSR / Russia had failed to register 53; of 2 154 launches the US had failed to register 191; and of 187 launches the PRC had failed to register 60.[18]

The Registration Convention makes no provision for independent validation of registrations or for reporting of observations that detect unregistered satellites unless they have caused damage to another space object or may be of a hazardous or deleterious nature. Other unresolved issues include whether any legal liability could be created if an item of debris attributable to one State destroyed, in outer space, an unregistered satellite belonging to another State, or the implications of an unregistered satellite damaging a registered satellite. In this context the provisions of the Liability Convention, already considered, might intersect with those of the Registration Convention.

Israel, which has indigenous space-launch capacities as well as satellites, has neither ratified nor acceded to the Registration Convention; neither have Egypt nor Malaysia, despite having satellites registered to them. A search on 29 January 2014 of the UN Registration Submissions data base revealed Israel as registering two satellites, Egypt one and Malaysia five.[19] Whether these States have voluntarily registered all of their satellites is not known.

The weight of available information implies the Registration Convention is a desirable concept but one not always adhered to, including by signatories to the Convention.

The *Agreement on the Rescue of Astronauts, the Return of Astronauts and the Return of Objects Launched into Outer Space* (Rescue Agreement (1968)) is based upon humanitarian considerations and in that regard is broadly analogous to an outer-space version of the *Geneva Conventions*. The text recognises obligations, in the case of an unintended landing by a populated spacecraft, to rescue, assist and return astronauts to their launching authorities together with the space object that landed.

While the Rescue Agreement imposes obligations to attempt the rescue of spacecraft crew, it does not require launching States to publicise the data and related information commonly necessary to safely undertake rescue missions, except in non-specific terminology. The Agreement states in part (Article 2);

> *'If assistance by the launching authority would help to affect a prompt rescue or would contribute substantially to the effectiveness of search and rescue operations, the launching authority shall cooperate with the Contracting Party with a view to the effective conduct of search and rescue operations.'*

At a minimum level, a State attempting or likely to attempt any such rescue would require *prior* advice on the radio frequencies proposed to be used by any specific spacecraft in an emergency, pavement length needed for a terrestrial landing if applicable, spacecraft thermal and chemical emission details with implications for the safety of rescuers, and guidance as to a means of entering the spacecraft without detonating explosive ejection devices.

Applicability of the Rescue Agreement in some circumstances might be influenced by the absence of a defined boundary between the terrestrial environments and the space region, under international law. If a receiving State did not regard a craft in its possession as a *space* vehicle then other laws, including domestic criminal laws relating to espionage, might be applied to both the vehicle and its occupants.

The *Agreement Governing the Activities of States on the Moon and other Celestial Bodies* (Moon Agreement (1984)) extended and clarified principles already endorsed by signatory States in the Outer Space Treaty. These included: a requirement that due regard be had for future generations and higher standards of living resulting from exploration and use of the Moon, increased emphasis on the need for international cooperation, a specific requirement to inform the Secretary-General of the UN about activities involving the Moon, approval of activities to collect and remove samples, the establishment of an international regime for exploration of the Moon, and clarification of requirements for the protection of people located on the Moon.

The Moon Agreement was an anathema in the US. Requirements in the Agreement for cooperative sharing and disclosure of activities appeared communistic as well as likely to discourage private venture capital needed for exploitation of resources on the Moon or on other celestial bodies. Within the US, organised lobby groups actively opposed adoption of the Agreement.[20] The potential for the Moon and other celestial bodies to contain vast and valuable resources was unknown at that time and it seemed imprudent to risk any avoidable constraints on their exploitation.[21]

The Moon Agreement was adopted in 1979. It then took until June 1984 to obtain ratification by the first five nations which allowed the Agreement to enter into force. No permanent member of the Security Council ratified or acceded to the Moon Agreement, although France signed it without taking binding action. Enthusiasm for the Moon Agreement might also be gauged in part from the terms of its Article 18, requiring that 10 years after it entered into force, viz. in June 1994, the possibility of a review be entered into the agenda of the General Assembly. In 1995 a General Assembly Resolution merely took note of a recommendation from the Committee On the Peaceful Uses of Outer Space (COPUOS) that no action be taken at that time to revise the Moon Agreement.[22]

The Moon Agreement eventually attracted minimal international support. As at January 2014, 19 UN member States had signed, ratified or acceded to it. Of the 19, only four States owned space launch facilities (France, India, Pakistan and Kazakhstan, although Kazakhstan's launch facility had been leased to Russia).

Resistance by States, in 1979 and subsequently, to detailed arrangements governing activities undertaken on the Moon offers an ironic example of the pervasive, subtle and enduring influence of international law. Indian and Chinese actions and intentions, among those of other States, to send missions to the Moon have caused alarm in the US regarding protection of its lunar artefacts from previous US missions.[23] On 20 July 2011 NASA released its *Recommendations to space-faring entities: how to protect and reserve the historic and scientific value of U.S. Government lunar artefacts,* containing detailed but voluntary guidelines for that purpose. On 8 July 2013, the *Apollo Lunar Landing Legacy Act* was introduced to the US House of Representatives with the objective of declaring a Historical Park in accordance with US law to protect those artefacts on the Moon primarily associated with landings under the US Apollo missions of the 1960s and 70s.[24] In November 2013, an article published in the journal *Science* advocated a bilateral agreement between Russia and the US, available for endorsement by other States, with a view to protecting lunar artefacts.[25] Of these initiatives, only the suggested bilateral agreement, if it eventually proceeds, will have credibility under international law and then it will apply only to States signatories to the extent they mutually accept and adopt it. Had the US committed to the Moon Agreement, several of its clauses would have provided a clear foundation for protecting its identified lunar sites.

The potential for mining and other forms of resource extraction on the Moon leads to additional and complicated issues associated with, but not totally dependent upon, the Moon Agreement. Considering numerous omissions and vagaries in the Moon Agreement as well as in the widely-accepted Outer Space Treaty, it has not been difficult for legal scholars to raise concerns many of which impliedly have potential to cause inter-State friction or even conflict. However, some leavening influences apply also.

Although surveys are incomplete, minimal evidence has emerged to indicate it will be viable to mine the Moon other than for scientific investigation, to manufacture fuel enabling a multi-stage journey elsewhere, or to support human sustenance in the lunar environment. In a longer term, probably beyond the 2030 CE sunset limitation on this investigation, it is conceivable the development of helium-based fusion reactors will create a demand for Helium-3 (^{3}He) which could be obtained on the Moon. But even if this fusion technology eventuates it is not clear that lunar acquisition of ^{3}He would be preferable to its synthesisation. Within the foreseeable future, aggressive competition for real estate on the 37 800 000 km^2 surface of the Moon seems unlikely, even allowing for the uneven geographic distribution of most known lunar resources.

Issues involving the Moon do not exist in a jurisprudential vacuum. The Outer Space Treaty extends to the Moon and other celestial bodies declaring them free for exploration and use by all States. It also addresses responsibility for non-government entities operating on the Moon and requires avoidance of lunar contamination. The terrestrial sale of lunar material, without attracting widespread challenge or complaint, apparently began in 1993 with implications for customary international law.[26]

If exploitation of lunar resources evolves to include commercial or other large-scale operations then a second and conceivably more comprehensive version of the Moon Agreement is likely to be drafted either as a treaty or as a bilateral or multilateral arrangement. Meanwhile if the Moon Agreement is a guide, States will be reluctant to endorse any new arrangement without comprehending the economic, technological and geopolitical environment in which it might operate.

Recognition of this emerging attitude provides an appropriate entrée to activities sometimes categorised as *soft* law.

The law of nations is often less precise than might be inferred from the above consideration of treaties and related arrangements. Situations arise in which two or more States will agree to: pursue or refrain from identified behaviours, apply mutually endorsed principles, or pursue agreed objectives although they are not legally bound to do so. These and similar circumstances are often referred to as soft law. With soft law, the extent of innate flexibility or commitment is commonly neither clear nor consistent, and can vary between arrangements, interpretations and specific circumstances.

For example, international negotiations which might lead eventually to a treaty-level agreement can provide an interim basis for adherence by one or more States to a specified behaviour. This occurred during the period 1964 – 67 when the US and USSR agreed to not place nuclear weapons in orbit pending the passage of an applicable treaty. A State might contribute to international stability by agreeing in principle to comply with a treaty despite not having ratified it, an example being the US generally complying with the *Convention on the Law of the Sea* without having ratified that convention.[27] Although these actions might or might not be categorised as international law, within an uncertain environment they contribute to predictability and can advance the spread of generally acceptable behaviours within the international community.

Soft law is becoming more prominent for issues related to outer space. The most obvious additional initiatives relating directly to outer space and adopted by the General Assembly were the five international principles, namely: *Principles Governing the Use by States of Artificial Earth Satellites for International Direct Television Broadcasting, Principles Relating to Remote Sensing of the Earth from Outer Space, Principles Relevant to the Use of Nuclear Power Sources in Outer Space,* and *A Declaration on International Cooperation in the Exploration and Use of Outer Space for the Benefit and in the Interests of all States, Taking into Particular Account the Needs of Developing Countries.*

In addition to those five principles, the *Hague Code of Conduct Against Ballistic Missile Proliferation* had attracted 137 subscribing States by 11 February 2014.[28] Some of the other international agreements of relevance to international security and with possible implications for outer space are listed in Chapter 7, Table 7 – 1. *Selected International Agreements; Agreements, Signatures, Ratifications, Accessions, Successions and Non-Party Status.*

Application of the Law

Within the context of aggression or warfare in outer space, the effective application of international law to sovereign States and to entities for which they are liable in outer space, would be both difficult and controversial. Not one of the five outer space treaties was formulated specifically to address warfare in space, although numerous provisions were intended to enhance global security, such as the ban against placing weapons of mass destruction in orbit. There has been broadly-based support for *The Hague Code of Conduct Against Ballistic Missile Proliferation*, but the Code was intended primarily to suppress proliferation. The most prominent States involved already had obtained their stocks of ballistic missiles.

In accordance with the UN Charter (Article 1), a purpose of the UN is to '*...take effective collective measures for the prevention and removal of threats to the peace, and for the suppression of acts of aggression or other breaches of the peace, and to bring about by peaceful means, and in conformity with the principles of justice and international law, adjustment or settlement of international disputes or situations which might lead to a breach of the peace.*' Intervention by the Security Council must be in accordance with guidelines in Chapters VI, VII, VIII and XII of the UN Charter although those provisions essentially follow Article 1. It seems possible, or probable, many actions able to be undertaken in space contrary to the Outer Space Treaty, or to other components of international law, would not be regarded as sufficiently serious to trigger a response by the Security Council. These actions might include: placing into orbit an unregistered satellite or refusing to compensate for damage caused to a third party by a space object.

Exacerbating challenges confronting the application of international law to outer space, is that some States members of the Security Council have not ratified, acceded to or signed specific space-related treaties. The following Table 6 – 2. *Signature, Ratification or Accession to UN Space Treaties by Members of the UN Security Council*, records the signature, ratification or accession to each space-related treaty by the States which were members of the Security Council in January 2014. I also undertook additional reviews, not included here, for dates when State membership of the Security Council was different to that of January 2014, but with similar results.

Table 6 – 2. Signature, Ratification, or Accession (Y/N) to Outer Space Treaties by Members of the UN Security Council (Compiled January 2014)[29]

State (Titles Abbreviated)	Outer Space Treaty	Rescue Agreement	Liability Convention	Registration Convention	Moon Agreement
US	Y	Y	Y	Y	N
Russia	Y	Y	Y	Y	N
PRC	Y	Y	Y	Y	N
Britain	Y	Y	Y	Y	N
France	Y	Y	Y	Y	Y
Argentina	Y	Y	Y	Y	N
Australia	Y	Y	Y	Y	Y
Chad	N	N	N	N	N
Chile	Y	Y	Y	Y	Y
Jordan	Y	Y	Y	N	N
Lithuania	N	N	N	Y	N
Luxembourg	Y	Y	Y	N	N
Nigeria	Y	Y	Y	Y	N
Republic of Korea	Y	Y	Y	Y	N
Rwanda	Y	Y	Y	N	N

From Table 6 – 2 it is apparent two of the 15 members of the Security Council had not signed, ratified or acceded to the Outer Space Treaty, one Member State, Chad, was not a party to any of the five space-related treaties, while only three of the 15 had signed, ratified or acceded to the Moon Agreement.

Four of these treaty-status agreements had been either in part or substantively negotiated between the USSR and the US during the Cold War when both nations were competing for global influence.[30] Those four agreements (Outer Space Treaty, 1967; Rescue Agreement, 1968; Liability Convention, 1972; and Registration Convention, 1976) each contain obvious and avoidable areas of imprecision. The imprecision might have resulted from inadvertent oversights reflecting intense and complicated negotiations, or could reflect a desire by negotiating parties to create a jurisprudential veneer beneath which they could act with minimal constraint. In this context, it might be pertinent that the Moon Agreement (1984), which included clarifications of existing agreements, was unacceptable to most powerful States.

Several versions and variations of a *Prevention of Arms Race in Outer Space* (PAROS) treaty proposal have been debated within the UN and associated forums since 1981 with little or no substantive progress having been made. COPUOS, an organisation nominally at the epicentre of formulating the international law of space, has for more than 40 years, been attempting unsuccessfully to *define* the lower boundary of space.[31] For over a decade, the European Space Agency and European Union have been attempting to develop an outer space code of conduct aimed in part at reducing space debris.[32]

The numerous attempts to progress international law relating to outer space security were inferentially recognised by the UN which adopted a resolution on 8 December 2010 proposing, *inter alia*, establishment of a group of government experts to consider transparency and confidence building measures.[33] The group subsequently identified a number of such measures and recommended States implement them on a voluntary basis.[34] These voluntary, and thus non-legally binding, measures essentially provided guidelines for behaviours recognised by some States as desirable. At the same time they allowed for all States to circumvent or ignore them, if they decided to do so, while simultaneously repudiating claims that international law had been breached. Whether this initiative advanced or retarded outer space security is not yet clear.

It is difficult to avoid the conclusion that, under existing international arrangements governing the UN, its committees and its subordinate agencies, progress in developing international law to support peace and security in outer space will be at best slow and challenging if it occurs at all. This opens the potential for other organisational arrangements to emerge and challenge existing mechanisms governing the international law of outer space should dissatisfaction and frustration with extant processes increase.

The Antarctic Continent as a Precedent for Peace

The Antarctic continent has been referred to as a terrestrial region to be used for peaceful purposes only (*Antarctic Treaty*, Article 1). Therefore it might appear to offer an exception to the near omnipresence of terrestrial warfare and, by implication, a precedent for retaining all or some part of outer space as a legally endorsed region of peace. That inference may be based more on hope than on reality.

Negotiations leading to the *Antarctic Treaty* began in the late 1940s. By November 2011 a total of 50 States had become parties to the Treaty, including all five permanent members of the UN Security Council. Consequently, the Treaty had developed into an arrangement with global implications.

The *Antarctic Treaty* (Article IV), states nothing in the Treaty is to be interpreted as a renunciation of previous rights or claims to territorial sovereignty in the Antarctic. It continues by asserting no acts or activities while the Treaty is in force shall constitute a basis for territorial sovereignty in the Antarctic. In essence, all existing or prospective territorial claims were neither confirmed nor repudiated but instead became held in indefinite abeyance.

The concept of holding territorial claims in abeyance has been referred to by one author as '*the sovereignty time bomb*'.[35] Seven States: Australia, Britain, New Zealand, France, Norway, Chile and Argentina, claim areas of Antarctica. Three of the States, Britain, Chile and Argentina have overlapping claims. The US does not recognise any national claim to Antarctica but reserves the right to make its own claim in the future. Similarly, the PRC does not recognise territorial claims in Antarctica but is increasing its investment in, together with its presence on, the Antarctic Continent. Brazil has declared it has a '*zone of interest*' in Antarctica. More claimants to Antarctic Territory might yet emerge.

Some claimants to Antarctic Territory have continued to assert 'rights', the validity of which is unclear in international law, although those actions if maintained have potential to cause international tension. For example:

> *'In a 2008 case, involving whaling in waters adjacent to Australia's Antarctic Territory, the defendant claimed the relevant region was part of the high seas. An Australian court found that the defendant had contravened the provisions of the Australian* Environment Protection and Biodiversity Conservation Act 1999, *but did not (and likely could not) rule on the legality or otherwise of the allegedly offending whaling under international law.'*[36]

In other circumstances nationalistic attitudes may have been encouraged by the popular media. One published news article, among numerous others in a similar vein, claimed the former USSR, the PRC and India had established bases on Australian Antarctic Territory without seeking permission from Australia.[37] While the claim was apparently accurate, the States mentioned had no obligation to seek Australia's permission to establish bases. To have done so could have strengthened Australia's claimed Antarctic sovereignty potentially at the expense of their own prospective assertions. When I searched records maintained by the Secretariat of the Antarctic Treaty, they indicated Russia, India and the PRC appeared to have met all international obligations for notification of their Antarctic activities.[38] Australia is represented at the related Antarctic Treaty Consultative Meetings.

The *Antarctic Treaty* preserves the Antarctic for '*peaceful purposes*' (Article I). That notwithstanding, notions of an Antarctica rich in valuable minerals have persisted for decades giving rise to interest in and, conceivably, temptations to exploit the resources. Resulting from surveys using contemporary scientific techniques the revised reality of a continent with sparse commercial mineral reserves has emerged and so one component of possible strategic destabilisation has dissipated but not evaporated.[39] The global need to optimise a diminishing terrestrial stock of minerals will, over time, increase and so interest in mining the Antarctic, including by some States not parties to the *Antarctic Treaty*, might yet arise.

Additionally the Antarctic continent is a desirable strategic asset, one likely to attract the attention of any major State power. It offers a geologically solid foundation for many activities, including space surveillance and related pursuits which, it could be argued, are not contrary to the *Antarctic Treaty,* Article I, it having prohibited ' ... *any measure of a military nature, such as the establishment of military bases and fortifications, the carrying out of military manoeuvres, as well as the testing of any type of weapon.*'

Should strategic circumstances deteriorate to the point at which weapons were installed on the Antarctic continent, long-range missiles stationed close to the geographic South Pole could be swivelled to change azimuth and thus to threaten terrestrial targets in any longitudinal direction. The technological capacity required to operate missiles and other weapons in the environmentally challenging regions of outer space has been demonstrated and that capacity is likely to be transferable to the harsh terrestrial environments of the Antarctic. There is minimal doubt Antarctica could, from a technological perspective, be militarised.

Control over the Antarctic continent would itself be an expression of significant power and influence. The Antarctic continent has a land mass of ~7 000 000 km^2, its surface area increasing through freezing of the surrounding sea in winter to more than 30 000 000 km^2. Australia, by comparison has a land area of 7 692 000 km^2.

The region above the Antarctic continent has implications for security in outer space. Satellites in polar and related orbits pass over or close to the South Pole. The area above the Antarctic has a relatively higher potential than most others to host collisions involving space objects. The concentration of satellites provides what military tacticians term a *choke point*; a region in which the presence of potential adversaries is concentrated. It is therefore a potentially prime site for disruptive activity involving satellites.

Antarctica has not yet hosted a war. However, the combination of unresolved claims for sovereignty, the potential for new and additional sovereign claims to emerge, continuing competition to emphasise and consolidate extant territorial claims, the potential strategic and economic benefits associated with dominance of the Antarctic continent, and its geographic proximity to a strategically significant region of outer space, all combine to depict a region of potential instability. It is therefore misleading to portray the Antarctic as an oasis of peace except possibly in a retrospective context.

Imperfect Outcomes

In summary, international law is a fragile mechanism upon which to base expectations that warfare in outer space will be significantly discouraged or contained. A body of data and information points to the likelihood that, without profound change and development, minimal advances will be made in the evolution of international law to support enhanced security for all spacefaring and other States. In particular, processes within and sponsored by the UN are biased toward achieving outcomes preferred by permanent members of the Security Council. The extent to which their policies and ambitions are likely to be motivated by egalitarian considerations is implicitly considered in the chapters following.

Chapter 7. State Space Policies and Warfare

In this chapter I consider key elements in space-related policies of the more prominent space-faring States, with the aim of clarifying their attitudes toward warfare in space and space security more generally. Analyses following also provide additional background information for consideration of the potential strategies to contain space warfare, reviewed in Chapter 8. The dialogue does not fit neatly with any categorisation or theory of international relations and it was not designed to do so.

Identifying or analysing policies of States toward the international environment, including outer space, is a hazardous undertaking. Numerous frameworks have been proposed to identify tools for comprehending and categorising the external actions and attitudes of nation States. Prominent authors have included Marx, Carr, Morgenthau, Keohane and Nye, Burchill and Linklatter, Buzan, and Chomsky, each advocating diverse and contrasting approaches. [1]

Burchill and Linklatter (2005), for example, distilled international relations debates into six specific categories: dominant actors, dominant relationships, empirical issues, ethical issues, issues about the philosophy of the social sciences and the prospects for '*multidisciplinarity*'. In contrast, Chomsky (1996) concluded historical conditions were too complicated to support any coherent theory of international relations, a notion that may have prompted a strong response from US diplomat Kissinger. [2]

Those diverse approaches do not provide a coherent framework for analysing State policies regarding outer space. Possibly in response to that impediment, published analyses have typically defaulted to focussing upon the ways in which a specific State might best achieve its own space-related objectives, most commonly and unsurprisingly by constraining the ambitions of perceived competitors.

Global perspectives, conversations sensitive to the aspirations of all spacefaring States, and of the many smaller States, have been rare.

Despite this, the UN and its agencies attempt periodically and with limited success to fill the global strategic vacuum, in part by encouraging global space engagement initiatives, such as the Platform for Space-based Disaster Management and Emergency Response, the Regional Centres for Space Science and Technology Education, and through the Basic Science Technology Initiative.

Discussion following is focused on nation States categorised as already possessing or developing a capacity to engage in hostile activities in the outer space region. That categorisation is traceable to evidence regarding weapons being developed or possessed by States, as presented in earlier chapters. Here, I have not examined the western European States separately, but instead considered western Europe as a strategic entity.

Framework for Extraterritorial Policies and Activities

Regardless of difficulties in analysing or categorising international conduct, there have been consistent and observable trends in the development and implementation of external activities by most States over time, although to categorise them as a coherent theory or philosophy would be fanciful. In Chapter 1 I considered some enduring traits evident throughout the recorded history of human warfare. Within an ocean of uncertainty those behaviours offered a constrained region of predictability on which further and more detailed analysis could, with reservations, be based.

In that context, three themes applicable to and commonly dominant in international relations, are identifiable. They are: States aspire to expand influence beyond their geographic borders, States attempt to maximise this influence, and States react adversely to events which impede, or threaten to impede, achievement of the first two objectives.

Almost all States aspire to exert influence beyond their borders. In some instances they are motivated by perceived needs to secure trade or to enhance national security. On other occasions States also want to impose their political, cultural or religious ideals on other nations as well as to expand physical territory.

Not every attempt to exert international influence has involved territorial conquest, however. The Cold War in which the main protagonists held contrasting economic, political and theological perspectives, is one example. Switzerland projects international influence through a network of monetary and investment arrangements. The State of the Vatican City exerts international influence through religion. But the external ambitions of most States have commonly influenced physical territory either because of conquest or through subordination by other means. This drive to exert influence has been apparent in both agrarian and industrialised States and has included initiatives by States relatively weak in relation to their perceived competitors.

These quests for expanded State influence have found expression in a wide variety of incidents and justifications. Britain, particularly in its period of power as head of an empire, initiated military action against more than half of the world's States, including locations as diverse as Cuba and Iceland.[3] Following India's independence from Britain in 1947, Portugal maintained a colonial outpost, Goa, on the Indian sub-continent until 1961 when Indian forces 'liberated' Goa. From 1976 the resentful population of Bougainville Island began seeking international recognition and independence from Papua New Guinea, itself an emerging State, to which Bougainville had been conjoined. In 1982 Argentina invaded the Falklands Islands on grounds that geographic proximity supported a historical right of governance superior to that held by Britain. During the period 1952 – 1976 Iceland, population 312 000, and Britain were involved in numerous disputes, commonly referred to as the '*cod wars*'. In 2012 Iceland entered a dispute with the European Union regarding fishing rights relating to the Faroe Islands. The list of such activities can be readily expanded but not completed because examples of extant or emerging States attempting to increase their influence continue to arise.

States not only attempt to *exert* international influence but also to *maximise* it, commonly to the extent their resources and political circumstances will allow. Foreign policies are therefore pursued within a competitive environment with the external interests of States overlapping as if components of an unstable Venn diagram.

Chapter 1 recorded examples of ancient empires and their subsequent downfalls. More recently, during the 17th to 19th centuries, several European States acquired and lost empires based on colonial territories located thousands of kilometres distant. In the 20th century the USSR through its Communist International (COMINTERN) initiative attempted to proselytise communist doctrine globally. Also in the 20th century, Germany and Japan tried to expand their influence across Europe and Asia respectively at first by conquest and later more successfully through trade. The US has on several occasions moved to reinforce its influence in the Caribbean region, on the South American continent and in other regions. Occasionally the competition for influence has been multi-tiered; for example, the PRC, Taiwan, Vietnam, Malaysia, The Philippines and Brunei all have territorial claims affecting the Spratley Islands.

Some competition has been camouflaged but not extinguished by treaties. The Antarctic continent is subject to the *Antarctic Treaty* which suspended all claims to territorial sovereignty in the Antarctic region. But the arrangement did not negate any competing, overlapping or pending territorial claims.

The desire to maximise influence becomes more obvious when powerful nations are involved and other States are collaterally affected. It is particularly destabilising when a new global power is emerging or attempting to emerge. This occurred in the 20th century when the US, accustomed to steadily increasing its global influence, encountered an ambitious and industrialising USSR. The resulting conflict arguably started in 1917 but became most prominent following the Second World War.

States resent obstacles to the achievement of these foreign policy objectives and their resentment is often palpable. Within the 20th century Germany's discontent over the terms of the *Treaty of Versailles* was one cause, conceivably the main cause, of the Second World War. The reactions of Middle-Eastern States to the creation of the modern State of Israel have resulted in several conflicts.

The Vietnam War occurred primarily because the US perceived a threat to its policy of containing Communist expansion. Both Turkey and Greece claim influence over Cyprus and its potentially resource-rich continental shelf, although the original grounds for their disputation are so ancient and complicated as to have become almost irrelevant. In 2012 Israel, perceiving itself to be the only nuclear-armed State permitted in the Middle East, threatened a pre-emptive strike against Iran which at that time had refused to allow international inspections of Iranian nuclear-related facilities.

Not all expressions of frustration lead to overt warfare or to the threat of it. Canada, for example, is described by the US Department of State as having extensive mutual defence interests with the US.[4] However, the US with a population nine times that of Canada's and a Gross Domestic Product (GDP) 12 times larger is the more powerful partner. Regardless of US power, Canada has consistently attempted to assert its perceived rights, leading to disputes concerning water, fishing, borders, protection of US conscripted personnel from involvement in the Vietnam War, and trade. Canada began processing nickel mined in Cuba when the US had in place a general prohibition on exports to or imports from Cuba. Australia, which views itself as an agent of influence in the Pacific region, interpreted a 2006 coup in Fiji as a potentially destabilising influence. Australia reacted with travel restrictions against Fijian officials and a modified foreign aid program.

Many other examples are available of reactions by States confronting impediments to their extraterritorial influence. Those reactions, combined with ambitions to expand and maximise international influence, provide a consistent background to the competitive tensions apparent in international affairs.

Concerning space, as I indicated in previous chapters, powerful States in particular have commonly viewed space as yet another environment into which they could extend their extraterritorial policies, attitudes and ambitions.

The above description of common and conventional international behaviour provides background to the following analyses of factors often claimed to underpin concepts of comparative international conduct.

Acceptance of International Law

Theoretically, records of State accessions to space and warfare-related treaties and similar agreements might indicate whether the framework of international law is likely to influence a given State's foreign policies in relation to outer space. But the key-word is *theoretically*.

Consistent refusals to become a party to normative international arrangements can indicate a State has consciously reserved the option of non-compliant behaviour. However, *acceptances* of treaties and similar arrangements are not reliable indicators of future compliance with them. Formal concurrence might demonstrate an intention to comply with both the letter and the broad intent of an agreement. It might also reflect an act of political convenience; one intended to provide a facade for the exploitation of exclusion clauses and loopholes within the international law.

Details of the selected States and their accessions to some international arrangements are shown in Table 7 – 1, *Selected International Agreements.* All forms of acceptance or endorsement including signatures without ratification are displayed as a composite 'yes' category (although mere domestic statements that a State will voluntarily comply with an agreement or protocol have not been included in that category). Consequently, this process does not reflect jurisprudential finesse.

Table 7 – 1. Selected International Agreements (Signatures [without ratification], Ratifications, Accessions and Successions (indicated collectively by 'Y') and Non-Party Status (indicated by 'N')) (Compiled January 2014) [5]

Agreement	State											
	US	Brazil	Britain	PRC	DPRK	India	Iran	Israel	Japan	Pakistan	Russia	France
UN Charter	Y	Y	Y	Y	Y	Y	Y	Y	Y	Y	Y	Y
Outer Space Treaty	Y	Y	Y	Y	Y	Y	Y	Y	Y	Y	Y	Y
Rescue Agreement	Y	Y	Y	Y	N	Y	Y	Y	Y	Y	Y	Y
Liability Convention	Y	Y	Y	Y	N	Y	Y	Y	Y	Y	Y	Y
Registration Convention	Y	Y	Y	Y	Y	Y	Y	N	Y	Y	Y	Y
Moon Agreement	N	N	N	N	N	Y	N	N	N	Y	N	Y
Non-Proliferation Treaty	Y	Y	Y	Y	Y	N	Y	N	Y	N	Y	Y
Missile Technology Control Regime	Y	Y	Y	N	N	N	N	N	Y	N	Y	Y
Comprehensive Nuclear Test Ban Treaty	Y	Y	Y	Y	N	N	Y	Y	Y	N	Y	Y
Code of Conduct Against Ballistic Missile Proliferation	Y	N	Y	N	N	N	N	N	Y	N	Y	Y
Biological Weapons Convention	Y	Y	Y	Y	Y	Y	Y	N	Y	Y	Y	Y
Chemical Weapons Convention	Y	Y	Y	Y	N	Y	Y	Y	Y	Y	Y	Y
Conventional Weapons Convention	Y	Y	Y	Y	N	Y	N	Y	Y	Y	Y	Y
Anti-Personnel Landmines Convention	N	Y	Y	N	N	N	N	N	Y	N	N	Y
Convention on Cluster Munitions	N	N	Y	N	N	N	N	N	Y	N	N	Y
Rome Statute (International Criminal Court)	Y[6]	Y	Y	N	N	N	Y	Y	Y	N	Y	Y
TOTAL	**13**	**13**	**15**	**10**	**5**	**9**	**10**	**8**	**15**	**9**	**13**	**16**

Table 7 - 1 indicates most space-faring States listed have subscribed to the first four of the five space-related agreements, but there are exceptions. The DPRK is not a signatory to the Rescue Agreement or to the Liability Convention. Israel is not a party to the Registration Convention. Additionally, the refusal of all States identified, except France, India and Pakistan, to subscribe in any form to the Moon Agreement might indicate reluctance to accept enhancements to, or clarification of, treaty-level international law affecting outer space.

Three of the unofficial nuclear-weapon States (viz. States not acknowledged as holding nuclear weapons in accordance with the Non-Proliferation Treaty but nevertheless known to have stocks of such weapons) notably India, Pakistan, and Israel, have predictably not signed the Non Proliferation Treaty, although the DPRK has done so. India, Pakistan, and the DPRK, have not signed the Comprehensive NTB Treaty either, although Israel has done so. Of entities listed, only five States (Russia, the US, France, Britain and Japan) concurred with the *Hague Code of Conduct Against Ballistic Missile Proliferation*.

Regarding agreements designed to limit specific non-nuclear weapons (biological weapons, chemical weapons, certain conventional weapons, landmines and cluster munitions) only Britain, France and Japan subscribed to all five arrangements.

Of the total sixteen arrangements listed, the DPRK committed to five and Israel to eight. In contrast, France committed in some form to all sixteen with Britain and Japan each committing to fifteen. The US and Russia each committed to thirteen and the PRC and Iran each to ten.

If this broadly based analysis could be employed to categorise States in accordance with their notional intention to comply with international law, then France, Britain and Japan would appear to be the most prospectively compliant States, the US and Russia together possibly with the PRC and Iran, would reside in a middle tier. The DPRK, India, Israel and Pakistan would seem least likely to be compliant.

However, impressions of resistance or compliance, while possibly indicative must be regarded with caution. First, the implications of accessions and of signatures without confirmation are imprecise. Second, some States have negotiated extensive networks of bilateral agreements which might have contributed to international stability, and so their propensity toward compliance could be higher than suggested.[7] Third, not all such arrangements can be considered equally significant. Repudiation of the Rescue Agreement, for example, would have consequences only in the historically unlikely event of a non-compliant State having an opportunity to rescue astronauts or their spacecraft. Additionally, in some instances technological developments can erode the value of extant treaties.[8]

Consequently, it is not possible to gain more than a generalised impression of actual or potential compliance with international law from a summary of this nature except perhaps at the extremes where France, Japan and Britain seem more likely to respect international law than is, for example, the DPRK. A separate and more detailed study is required to investigate further the nexus between acceptance of treaties or related instruments and respect for international law.

Histories and Space Policies of Space-Faring States

With one exception, the strategic and related histories of selected space-faring States identified in Table 7 – 1 are now examined in an attempt to indicate their past, and imply future, approaches to space security. The exception is that France has been included under the generic heading of *Europe*.

US

Soon after its *Declaration of Independence* in 1776 the US began a rise toward becoming a globally influential State. During President Jefferson's period in office (1801 – 1809) the US land area was approximately 2 000 000 km^2. Following expansion as a consequence of warfare, barter and purchase, it increased to approximately 9 000 000 km^2. The US also acquired several external territories, holding them for varying periods of time.

The US has been exerting significant international influence for almost two centuries. One early indication of its expanding imprint was the *Monroe Doctrine* (1823) which effectively warned European nations the US would not tolerate interference in the North or South American continents, a claim reinforced in the 20th century during negotiations underlying the *Covenant of the League of Nations*.[9]

In 1945 the US unilaterally asserted rights to undersea resources located on its continental shelf, leaving other States to imitate the precedent if they wished.[10] During 1963 the US determined it had '*adequate authority*' to penetrate the territorial waters of the PRC, DPRK or Cuba to retrieve Gemini personnel or spacecraft if the need to do so arose.[11] The US, among other States, has also published unilateral declarations of exclusion zones over the international high seas for purposes of weapons testing.[12] When the USSR attempted to broaden its international influence, particularly following the Second World War, the US responded and the Cold War began.

Although it has not conquered and retained large tracts of land, as at 2014 the US still held nine overseas territories.[13] The US had also established military sites which encircled Earth, numbering at least 550 installations.[14]

The US maintains diplomatic relations with States throughout the world, commonly with multiple missions in single States. So far as could be determined the US has the most comprehensive network of bilateral and multilateral treaties of any nation State. As at 2011 it had bilateral treaties with 232 nation States and their colonies.

US international liaison includes obvious elements such as its close relationships with Canada, Britain and Japan as well as arrangements rarely publicised. The US has had for example a long-established commercial arrangement with Russia which supplied RD-180 rocket engines to the US for its Atlas launch vehicles. In 2010, nuclear reactors and related machinery were among US exports to the PRC.[15]

The US has been at war for approximately 10 per cent of its existence. During the 20th century it established a trend toward increasingly frequent military action, being at war for 15 % of the time, and in the second half of that century for 22 %. In the 21st century, from 2001 until 2014, the United States was constantly at war.

US space policy reflects that background. It was originally influenced by a perceived need to counter threats posed by rocketry developments within the USSR and an intention to match or exceed that capability. Competition between the US Air Force, Army and Navy then prevailed, the only common theme being that activities involving outer space were perceived as military and national security issues. This link between US national security and outer space was in part entrenched by the Strategic Missile Evaluation Committee, led by von Neumann. In 1954 it recommended acceleration of a ballistic missile program. That finding supported the US Central Intelligence Agency's ambition to gain high-altitude intelligence collection capabilities through launches into space.[16]

The US scientific community became observably involved with space activities through a proposal to launch a satellite as part of the planned International Geophysical Year (1957). At that time the US National Security Council noted the potential of this initiative to assist with anti-missile research, as well as recording the assessment that '*a satellite would constitute no active military offensive threat to any country over which it might pass.*'[17]

On 18 August 1958 the National Security Council prepared a *Statement of preliminary U.S. policy on outer space*. It included the objective of, '*...achieve a military capability in outer space sufficient to assure the overall superiority of US ... offensive and defensive systems relative to those of the USSR.*'[18] In 1959 the National Security Council proposed a draft policy including an intention to study, '*passive and active defense systems to detect and destroy enemy missiles or space vehicles....*'[19] The 1959 document apparently was amended and formed the basis of a space policy released as a National Aeronautics and Space Council paper on 26 January 1960 but I did not find a copy of the January 1960 document.

Under President Carter, the US national space policy, 1978, stated the US space program would be conducted in accordance with the principle that exploration and use of outer space would support the '*national well being and the policies of the United States.*' Subsequently, President Reagan's space policy, 1988, mentioned space control, requiring the US Department of Defense to:

> '... *develop, operate and maintain enduring space systems to ensure its freedom of action in space. This requires an integrated combination of antisatellite, survivability and surveillance capabilities.*'

Space policies authorised by Presidents GHW Bush and WJ Clinton followed the lead given by President Reagan, with President Clinton's policy incorporating a requirement to '*maintain the capacity to execute the mission areas of space support, force enhancement, space control and force application.*' The 2006 space policy authorised by then President Bush was similar but more proactive. It included deterring others from developing capabilities intended to impede US rights, developing the capacity to ensure freedom of US operations in space and to deny adversaries the use of space capabilities hostile to US national interests. The US was authorised to use nuclear power systems, undefined, for space exploration or operational capabilities.

President Obama's space policy, 2010, retained the same elements of ensuring freedom of action to the US and denying it to others if necessary. Nuclear options were retained also.[20]

While optimising its own domestic flexibility, the US has pursued a policy of constraints upon the activities of some other States. From Table 7 - 1 it is clear the US has endorsed in various ways the Non-Proliferation Treaty, Missile Technology Control Regime, Comprehensive NTB Treaty, and Code Against Ballistic Missile Proliferation. US export controls include aspects of space-related technologies.[21] These controls are regularly reviewed and can be supplemented by additional arrangements. In 2009 India and the US entered a Technology Safeguards Agreement. As at February 2014 the US had also signed bilateral space awareness agreements with France, Australia, Japan, Italy and Canada.

From previous discourse it is obvious the US has during five decades been developing and testing weapons for potential use in outer space. It now has capacity to detonate nuclear weapons in space, conduct kinetic and related attacks against satellites and apply advanced electronic warfare technologies. The US is also one of the leading States, possibly *the* leading State, in developing space planes and hypervelocity engines. It experiments with new materials and technologies, employing in part a network of State-owned and well-credentialed university laboratories to conduct research.

Successive US governments have claimed the US has global leadership in outer space but the concepts, definitions, verifiable data and other evidence to support this ambitious assertion all await clarification, an issue further considered in Chapter 8.

Given its military history and long-established policies, offensive capabilities and its investment in space weaponry the US is as prepared as any other State to conduct warfare in outer space.

Brazil

After gaining independence from Portugal in 1822, Brazil in its various political manifestations emerged victorious from every external conflict in which it became engaged, including the First and Second World Wars when it supported the allied causes. In contrast, over the same period Brazil was domestically unstable having experienced numerous internal revolts and other disruptions. This might in part be attributable to Brazilian ethnicity which reflects a complicated national history resulting in many Brazilians sharing elements of European, American-Indian, Asian, African and Middle Eastern origins.

Relative political stability emerged during 1985, with the electoral-college election of a president. Direct elections began in 1989. By January 2014 the Brazilian National Congress still included representatives of 12 different political parties.

Brazil has pursued a security policy emphasising its independence. This has likely given the State more influence and created better opportunities for increased policy flexibility than would have been achieved by any alternative strategy. Given its background and policy it was almost inevitable Brazil would attempt to build a national space industry

In doing so Brazil has established a wide-ranging space-related relationship with the European Union including the negotiation of arrangements relating to science, and communications. Brazil also cooperates over space-related issues with States as diverse as Canada, the PRC, Russia, France and Ukraine. Argentina and Brazil have fostered a close relationship. The US, also, has pursued constructive interactions with Brazil in matters affecting outer space, despite occasional concerns regarding the implications of Brazil's non-aligned strategic policy. Brazil and the PRC have cooperated over the construction and launch of several satellites.

The first Brazilian satellite was launched on 9 February 1993 using a US launch vehicle. Brazil's indigenous launcher, the VSV-30 has conceivably not met expectations, although it reached an altitude of 250 km on 24 October 2004 (the Cajana Test) with a payload of ~ 400 kg. Russia and Ukraine have been both competing and cooperating to provide more substantial launch vehicles to Brazil. One, Ukraine's Cyclone 4, had scheduled a test launch for 2015.[22] [23] However, I could not determine the implications for these arrangements of a major confrontation between Russia and Ukraine during 2014 and continuing.

Brazil's National Strategy of Defense includes reference to three strategic sectors as *'essential for the national defense'*, namely *'cybernetics, space and nuclear'*. The document repudiates any intention to build nuclear weapons but establishes as a priority, *'Design and manufacture satellite-launching vehicles and develop remote guiding technologies, especially inertial systems and liquid fuel propulsion technologies.'*[24]

Generalised guidance as to Brazilian space policy was found in the legislation that created Agência Espacial Brasileira (Brazilian Space Agency). Brazilian Law 8.854 delegates to that Agency responsibility for advice and implementation of higher-level guidance with emphasis on rationalisation of national assets, research and education.

The Agência Espacial Brasileira appears focused on promoting autonomy in space although this will be difficult or impossible to achieve in the foreseeable future. A published presentation by e Silva, Liporace and Santos (2007) described Brazilian activity across a broad spectrum of space activities and applications, including launch, supported by a network of almost global liaison on space issues.[25]

In 2008 Agência Espacial Brasileira released a summary of its activities and objectives relating to space, *Space, it's about people*. The document listed Brazil's industrial, economic and intellectual investment in space-related initiatives. National security was not mentioned. In June 2011 a joint announcement from the Ministry of Science Technology and Innovation and Agência Espacial Brasileira advised a new space policy focussed on developing satellites with corporate participation. On 25 October 2011, a news release advised the space agency's financial budget would be almost doubled to address a range of requirements including satellites and launch vehicles. No indications of a change in strategic direction for national space engagement were apparent at that time.

However, during early 2013 Brazil again reviewed its outer space program, scaling back or cancelling several launch-related initiatives although development of the Cyclone 4 was preserved. Further confirmation of this change appeared when, in January 2014, the US company Comtech Telecommunications Inc. was awarded a contract to upgrade the Brazilian satellite network instead of the work being allocated to a Brazilian organisation.[26] The strategic implications of this development have not yet matured (2015).

The economic viability of Brazil's Alcântara space-launch site depends upon its utilisation by other States, implying Brazil would be unlikely to risk threatening or alienating them and potentially losing customers.

These considerations, together with Brazil's non-aligned international posture, as reflected in both its breadth of international relationships and its adopted space policies, point congruently toward Brazil not developing a space warfare capacity within the foreseeable future although it could acquire the launch and nuclear technologies to do so. The national security aspects of Brazil's space access therefore seem likely to be focused upon passive data collection and subsequent data processing. No indication was found that Brazil was likely to change its non-aligned security posture.

The possibility Brazil might eventually acquire its own launch vehicles nevertheless remains a cause of concern for the US, partly because of the potential for those vehicles to be used as missiles and partly because any such missile might in the future be coupled to a nuclear weapon, given Brazil has an advanced nuclear-energy program.

Britain

Following centuries of invasions, fragmentation and disruption, by the mid 17th century Britain, England in particular, was stabilising as an enduring entity it having become more cohesive than formerly. Britain later adapted to and at times led the political and economic transformations common to much of western Europe including an increased role for legislatures, the industrial revolution, and voting rights for adults. Unlike most modern nations, Britain did not develop and still does not have a written national constitution.

The First and Second World Wars imposed severe financial stress on Britain. Following the Second World War it maintained a veneer of prosperity, but its colonial empire effectively had dissolved. Industrial unrest had increased as did other sources of disruption, including a long-running and complicated relationship involving politics in Ireland. Nevertheless Britain, in part because of its prominent role in the Second World War, retained prestige sufficient to allow its acceptance as a permanent member of the United Nations Security Council, giving it significant international influence. Britain has also developed a comprehensive network of global diplomatic connections.[27]

Despite setbacks, during the 1950s Britain bolstered its strategic profile in part by designing and testing a nuclear weapon. It also tried to develop an indigenous long-range missile but failed to persist with it. Britain therefore did not acquire either its own strategic missile or an integrated domestic space launch capacity.

Available evidence points to Britain's defence and related national security relationships being focused on the US and its allies together with an occasionally ambivalent engagement involving Europe. Reflecting this bifurcation of strategic priorities Britain joined the European Space Research Organisation which subsequently merged with the European Launcher Development Organisation to form the European

Space Agency. Simultaneously, Britain cooperated with the US on a range of unspecified space defence initiatives with the two States negotiating cooperative defence and economic arrangements.[28] On 15 February 1960 Britain agreed to host a US ballistic missile early warning station at Fylingdales Moor.

Britain possesses a nuclear warhead arsenal totalling approximately 225 warheads.[29] Details regarding the range of weapons to which these warheads could be attached have not been released. However, through nuclear-weapon-armed Trident missiles, Britain apparently acquired a strategic weapon with a range of 7 400 – 12 000 km indicating the trajectory could enter into outer space. But the missiles were purchased from and serviced by the US. It is not clear if Britain has or could gain from the US approval to operate the weapon unilaterally in outer space or elsewhere.

A capacity to design and build satellites and related sensors in Britain has been established substantially by commercial and academic organisations with links to Britain's Ministry of Defence. Conceivably, its military satellites are equipped with some forms of electronic defensive capabilities.

A space policy and oversight organisation, the British National Space Council, later renamed UK Space Agency, was formed in 1985. A subsequent (2012) policy statement, *UK Civil Space Strategy 2012 -2016*, was focussed on development of a civil space industry.

I did not find a formal British military or security space policy. However, because 90 per cent of British military platforms and systems within the military equipment program are reliant on space to some degree a classified national space policy almost certainly exists. [30] In a speech delivered on 12 July 2010, the Assistant Chief of the [British] Air Staff mentioned a document, *2008 Space Security Policy.* [31] Deductively, the policy must incorporate reference to management of Britain's military communications satellite constellation, the Skynet, as well as the installation at Fylingdales which now tracks 3 000 objects in orbit. Britain amalgamates data from Fylingdales and other sources to construct a recognised space picture including detail of overflights by foreign satellites.

Available information implies Britain's capacity to engage directly in space warfare is and may remain minimal except through the possible application of foreign designed and maintained missiles. However its participation in space observation, including monitoring the overflight of foreign missiles, indicates a strong linkage to US military arrangements concerning outer space. That aside, no evidence was found suggesting Britain views national participation in space warfare as a priority.

PRC

The PRC has an enduring civilisation of ancient origin, the Chinese having achieved early advances in engineering, art, economics, astronomy, other sciences and military tactics. Regions of China were conquered during several millennia, but pervasive Chinese culture absorbed aggressors into the mainstream of its civilisation.

The absorption process did not function successfully following China's first major and direct association with the West. In the mid 19th century European nations and the US were demanding trading rights. Western States supported their demands on at least one occasion through an armed conflict, the Opium Wars, which erupted when China attempted to enforce its domestic ban on opium imports. In response, China was forced by predominantly British intervention to accept the drug as a trading commodity. The British intervention was supplemented by international attempts to convert Chinese people to Christianity.

Further anti-Chinese activity followed. During the Versailles negotiations following the First World War, concessions in Shandong Province were transferred from German to Japanese control. The US refused to abandon immunity from Chinese law for US citizens in China until the 1940s. In 1949 the US rejected the legitimacy of the Chinese communist government together with the formation of the PRC and did not extend formal recognition of the PRC until 1979. The US also supported the Republic of China (Taiwan) as a permanent member of the United Nations Security Council, in preference to the PRC, until that arrangement was overturned by General Assembly Resolution 2758 of 25 October 1971. Cold War tensions further affected the PRC's relationship with the West.

When the PRC adopted an 'open door' policy (c. 1978) some normalisation of relations with the US followed as did an enhanced relationship between the PRC and Japan, Hong Kong, Taiwan and South Korea, together with other Asian and European States. Antagonism between the PRC and the US reduced to some extent when the USSR began to threaten PRC security and US President Nixon visited the PRC in 1972.

Despite increasingly strong economic ties, the PRC-West relationship has often been cordial but reserved. Western sources continue to castigate the PRC for the Tiananmen Square incident of 1989, in which protesting students were killed by PRC government forces. More recently, in 2011, several Western States expressed concern when the PRC launched its first aircraft carrier. Objections again arose in 2013 when the PRC declared an Air Defence Information Zone which included air space above the Senkaku/Diaoyu islands, claimed as sovereign territory by both Japan and the PRC. The PRC's international trade, with implications for economic interdependencies, has nevertheless reached significant levels.

From the PRC's current perspective, it is geographically and politically surrounded by competitors with potential to become opponents. Russia, following the loss of its Asian empire, has moved both geographically and politically toward the West albeit haltingly and with constraints and reservations, some of which appear to be increasing. Japan is viewed by the PRC as aligned with the US, as is the RoK. An increasingly powerful India is emerging to the south. The DPRK is possibly perceived more as a source of potential destabilisation than a constructive ally. US-encouraged and enabled missile and other defence arrangements are resulting in the partial encirclement of the PRC although the US claims this to be merely a coincidence.[32]

In essence, Chinese experience when interacting with Western nations has often been negative. That legacy has provided minimal motivation for the PRC to seek a cooperative international security relationship with the West other than to promote its own economic objectives or to expand its own sphere of influence.

The PRC has a space agency, China National Space Administration (CNSA), through which it is expanding Chinese influence in global space activities. The CNSA has a website <http://www.cnsa.gov.cn> containing links to US, European, Russian and Canadian sites. Liaison with Spain and Brazil is also mentioned on the website as is the PRC's initiative in hosting the 2008 meeting of the Asia-Pacific Space Co-operation Organisation, membership of which includes Bangladesh, Iran, Mongolia, Pakistan, Peru and Thailand. The PRC has signed cooperative agreements with the European Space Agency, Brazil, France, Britain and Germany. In 2010 the PRC also signed a contract to build a satellite for Bolivia, the *Tupac Katari* satellite.

The PRC's policy paper on space, *China's Space Activities*, 2003, neither specifically mentioned nor excluded space warfare from potential State objectives. It stated;

> *'China will continue to promote the development of its space industry in the light of its national situation, and make due contributions to the peaceful use of outer space, and to the civilization and progress of mankind.'*[33]

The remainder of the document maintained this ambivalence by including notions such as, '... *the exploration and utilization of outer space should be for peaceful purposes*', those words being positioned almost adjacent to, '*The aims and principles of China's space activities are determined by their important status and function in protecting China's national interests and build up the comprehensive national strength.*'[34] A subsequent PRC policy document, released in 2011, emphasised a commitment to the peaceful use of outer space but incorporated references to national security, national rights and interests and building '*national comprehensive strength*' these objectives having been recorded without further elaboration.[35]

The PRC apparently intends to pursue a range of space-related activities, some directed toward peaceful pursuits and others to national security. [36] A few initiatives imply dual-purpose roles in which commercial missions are combined with national security applications. The PRC's indigenous satellite navigation service, the Beidou, is one example of this duality.

The Chinese concept of normative space-related behaviour appears to be still evolving. On 11 or 12 January 2007 the PRC adversely affected the international space community by shooting-down its own satellite and creating more than 2 000 items of debris 10 cm or greater in size. On 13 March 2012, PRC lunar exploration official and deputy to the National People's Congress, Hu Hao, was reported as proposing a domestic space law, in part to protect PRC rights in outer space.[37] This impliedly would result in a document similar in content to US national space policies, given both States claim to function within the same basic framework of international law.

From previous chapters it is apparent the PRC is developing a comprehensive suite of outer space weapons. It has a history of possessing nuclear weapons together with the capacity to launch them into outer space. The PRC has twice successfully tested a weapon based on kinetic or conventional explosive forces and capable of deployment in an anti-satellite role. A successful docking between the *Shenzhou* space capsule and a module of the Chinese *Tiangong* space station on 18 June 2012 indicated an alibility to operate co-orbital anti-satellite technology with high precision. Given its comprehensive technological achievements, the PRC almost certainly could design anti-satellite electronic technologies as well as construct space planes, hypervelocity vehicles and micro-satellites.

Evidence that the PRC is developing space-based beamed-energy weapons in any form is scant and stems from claims by Western nations, those claims not having been accompanied by verifiable evidence. Given the US's apparently minimal return on investment in these technologies the PRC might have decided to either ignore such weapons or to rely upon espionage or data leakage if unanticipated developments arise.

The US and other Western nations are responding ambivalently to the PRC's space-related developments, being both concerned over implications for their own security and also intrigued by the possibility of exploiting commercial opportunities.

Future Chinese influence over the space region could develop in any of several ways with available options reflecting in part the PRC's expansionist ambitions as well as taking account of Western innovations and Western responses to PRC strategic initiatives. Given its past experiences in East-West relations and its emerging international status the PRC is unlikely to retreat in the face of future assertiveness or threats of aggression from any Western State.

One asset unique to the PRC in relation to the development of space technologies, is its demographic advantage of having the youngest cadre of space scientists and engineers of any major space power. Its members are on average two decades younger than equivalent US or Russian personnel.[38]

DPRK

The racially and linguistically homogenous Koreans have been invaded and controlled by external forces several times including twice by Japan. The most recent Japanese invasion resulted in the occupation of Korea from 1910 to 1945. Following Japan's surrender, ending the Second World War in the Pacific and Asian regions, its occupation of Korea ended and Korea became divided into two zones with the US administering the southern zone and the USSR the northern zone. In the north the DPRK came into being in 1948 with a premier, Kim Il-sung, in control. A communist political and economic philosophy was established. The DPRK invaded the RoK in 1950 but was repelled and an armistice signed in July 1953.

The DPRK's principal allies, to the extent it has allies, remain Russia and the PRC. Anti-Japanese resentment has persisted in the DPRK since Japan's surrender in 1945. The DPRK has maintained an often acrimonious relationship with the RoK.

The DPRK has experienced severe economic challenges, its estimated population per capita income being ~USD 1 800 per annum (2011 estimate). By comparison, the population per capita income of the RoK was, ~USD 33 200 per annum (2012 estimate).[39]

The DPRK's establishment of an indigenous nuclear weapons program has caused international consternation. The program has been condemned on multiple occasions by the United Nations Security Council. This notwithstanding, on 12 February 2013 the DPRK detonated its third nuclear test.[40]

DPRK has developed a space launch or ballistic missile capability despite the United Nations Security Council demanding it cease such activity.[41] The DPRK nevertheless persisted with attempts to launch missiles, or space-launch vehicles (the technological distinction between them being minimal) in July 2006, April 2009 and April 2012.[42] On 12 December 2012 the DPRK successfully launched an apparently small satellite into orbit.[43]

The official DPRK website describes its *Juche Idea* as the State's guiding principle. The Juche Idea is a generic concept embracing self reliance and self defence. Space-related issues appear to have been subordinated to this broader policy objective. No reference to a DPRK policy in relation to outer space was found. It is possible that the State does not have a space policy *per se*. Its claimed satellite launches may have been designed to camouflage a program of building long-range ballistic missiles.

The DPRK has pursued the Juche Idea apparently regardless of indigenous poverty. Nuclear-weapon developments have been funded despite the DPRK having a prior relationship with the PRC, itself a nuclear-armed State. Emphasising the DPRK's unorthodox policies, signs have emerged of an officially accepted supernatural faith in the continuing influence of a former leader, Kim Jong Il, who died in December 2011.[44]

In the context of global space security, the DPRK remains a potential threat. DPRK technology could eventually enable it to launch a nuclear weapon into outer space. It is difficult to identify any rational motivation for the DPRK to do so but that might not be an impediment. Treaties and related obligations under international law as accepted by other States are not apparently viewed by the DPRK as constraining national objectives and it has adopted relatively few of them.

Given the DPRK had numerous failures when attempting to launch its indigenous long-range rockets, and its only one success to December 2012, space weapons requiring precise guidance are apparently beyond its technological competence. They seem likely to remain so unless external support is provided. PRC and Russian support for Security Council resolutions condemning DPRK activities in relation to both nuclear weapons and launch vehicles or missiles suggest those States will limit their strategically significant assistance to the DPRK in the foreseeable future.

From the minimal and confusing information available it is not possible to confidently determine the DPRK's intentions in relation to warfare in the outer space region. Some evidence points to it eventually becoming a threat to international stability in that context.

India

Indian civilisation has been evolving over more than four millennia during which time northern regions of the present State were invaded several times. Small kingdoms formed and reformed. Islamic influences spread across the sub-continent but Hinduism also emerged and became influential.

British colonialism within India began in the 17th century and, having consolidated, continued until 1947. India then adopted a democratic model of government with an elected parliament, although national administration has at times been controversial and turbulent. Despite the 1947 separation of predominantly Hindu India from the mainly Muslim Pakistan, religious violence became and remained a socially destabilising internal and external influence.

India has adopted a foreign policy of non-alignment with the major powers or associated blocs. Following the 1947 separation, relations with Pakistan became fragile, particularly because Pakistan refused to recognise the state of Kashmir as a component of India.[45] India and the PRC have on occasions since been mutually suspicious, having for example fought a border war in 1962 without the basic disagreement being subsequently resolved. The PRC has been concerned over India's continuing relations with the USSR. In turn, India considers itself to have been threatened by the PRC's and DPRK's transfer of ballistic missile technology to Pakistan. These and related issues form the background to a border-related dialogue still continuing.[46]

The warmth of relations between India and the US has fluctuated over time. By the mid 1990s a closer interaction with US had emerged. However, Indian enthusiasm was diluted when the US began improving its relationship with Pakistan, conceivably as part of alliance arrangements designed to support the emerging battle against al-Qaeda. India was further concerned when the US appeared reluctant to pursue Pakistan for its India-alleged role in at least two terrorist-related incidents.[47]

The US nevertheless comprehends India's emerging influence and has been trying to establish a closer bilateral relationship. In December 2006 the US Congress passed the *United-States-India Peaceful Atomic Cooperation Act* allowing direct nuclear commerce with India. The US has also indicated it would support India becoming a permanent member of the UN Security Council.[48]

Regardless of diplomatic overtures from the US or elsewhere, India has maintained its non-aligned posture. In 2010 Russia and India signed an agreement for Russia to build

16 nuclear reactors in India. India has also fostered components of a more cooperative relationship with the PRC. But despite these diplomatic advances India has built a substantial military force of approximately 980 000 active troops, a Navy with at least 133 surface and 11 sub-surface vessels, a coastguard with some 93 vessels plus aircraft and an air force with approximately 1 500 aircraft.

India has been ambivalent, possibly deliberately so, regarding its policies related to outer space, in particular the militarisation of outer space. The Indian Space Research Organisation (ISRO) has highlighted India's focus on science, industry and the provision of services, offering agreements with 37 nations and multinational space organisations as evidence of an international and nonaligned posture.[49] The website of the Indian Defence Research and Development Organisation avoids reference to space issues. A comprehensive review of India's strategic posture and strategic planning, with emphasis on the nuclear component, omitted consideration of space warfare.[50]

India nevertheless has maintained missile programs which avoided most budget cuts during the 1990s. India is also patriotically proud of its nuclear weapons capability. A sudden resumption of nuclear testing in 1998 apparently reflected a desire by India to become an obvious and globally identifiable nuclear State.

India's space-warfare-related investments may have originated as components of a perceived need to contain the PRC, it being India's primary potential opponent. Despite improvements in the India-PRC relationship, the full impact of an industrialised PRC has not yet emerged, and the prospect of an expansionist PRC could from the Indian perspective justify a response.

From previous chapters it is apparent India is maintaining a program of preparing for possible conflict in space. In addition to developing missile-defence / launch vehicles, India has researched actual and potential anti-satellite technologies, including space planes and scramjets, as well as instigating research into a first-generation directed-energy weapon.

In keeping with its likely strategic environment India will probably maintain an investment in developing space warfare capabilities more for protection than aggression.

Iran

For at least 1 400 years Persia, now Iran, has been the site of frequent turbulence commonly as a result of either external interference or interventions in State affairs by religious devotees, both influences having persisted into the 21st century. Occasional support from Russia might have been based on Russian ambitions to limit US influence in the Middle East with Iran being merely a collateral pawn in a broader strategic framework. Assistance both to and from the international Moslem community could be a significant factor of influence in determining Iranian objectives and behaviours as well as providing access to finance but this could be almost invisible to conventional diplomatic observation. Analysis of Iranian-initiated external relations is thus challenging.

In 1941, when Shah Mohammad Reza Pahlavi assumed power, Iran became a substantially pro-Western State in which political interference by clergy was seldom tolerated. The cordiality of Iran's relationships with the US, Britain and the USSR varied but within the then existing Middle Eastern context the Shah of Iran was regarded by the West as a conservative leader.

Signifying significant national ambition, during c.1974 -5 the Shah signed agreements with the US to build eight nuclear reactors, with France to build five and Germany one.[51] Arms importation arrangements apparently included negotiations with the USSR for missiles and with Israel to secretly provide components of missiles denied to Iran by the US. By 31 December 2005 the International Atomic Energy Agency (IAEA) was still recording no nuclear reactors in Iran.

In 1979 the Shah was removed by a revolutionary force with its leader, the cleric Ayatollah Khomeini, proclaiming nationalist goals while being theologically and politically opposed to the US and to Israel. Khomeini seemed determined to export both oil and Islamic revolution. During 1995 Russia nevertheless announced it would proceed with the sale of nuclear reactors to Iran. Iran has since been castigated for breaching international nuclear inspection requirements with its non-compliance attracting international criticism, particularly from the US. As at 31 December 2012 the IAEA recorded Iran as operating one nuclear reactor.

In response to Iran's claimed non-compliance regarding nuclear facilities inspections, sanctions were imposed by the US, European Union (EU) and the UN Security Council. However, the restrictions might not have been effective in containing Iran's expanding missile inventory. In 2010 Iran reportedly obtained four S-300 Russian missiles, despite Russia's claimed refusal to deliver them. In May 2011 the US imposed sanctions against companies in Venezuela, Singapore, the PRC and Israel for dealing with Iran.

By early 2014 Iran had agreed to a *Joint Plan of Action* in which some sanctions would be eased if Iran allowed enhanced inspections of its nuclear facilities including enrichment facilities. The Plan did not affect existing limitations on weapons and related components. But successful implementation of the Plan might prove challenging. As at January 2014

Iran was, for example, defending its right to conduct research on advanced centrifuges which have implications for refinement of radioactive materials.[52]

Given its strategic history and geographic location, Iranian engagement with outer space is almost certain to be motivated by a combination of national defence requirements and the desire to project a more prominent international profile. Iran has a space agency, Sázmán e Fazái e Irán, formed initially to coordinate remote sensing and space-related communications. On 29 September 2010 the agency's mission was expanded to include oversight of all active Iranian participants in space-related activities. The agency now has direct responsibility to the Institution of the Iranian Presidency.

A news-link on the Iranian space agency website includes media releases regarding intentions to launch an animal into space, develop human space suits and to launch a large communications satellite, ambitions which cumulatively might imply an interest in science *per se* but could also overlay ambitions to develop or acquire more powerful launch vehicles, or missiles, than Iran presently has.[53]

Compared against established space-faring States such as the US, PRC and France, Iran is now at a relatively early stage in building its space technologies. Iran lacks the capacity to become a participant in hostile activity in outer space except, conceivably, through the use of lasers against imaging sensors carried on satellites. If the proposed nuclear facility inspection regime fails to prevent Iran developing nuclear weapons then Iran might eventually acquire the capacity to launch a nuclear weapon into outer space. However, Iran does not have a record of acting irrationally as judged by its own national ambitions and there is no obvious reason for it to commit such an act.

The extent and direction of Iran's space technologies is uncertain. Equally uncertain is whether sanctions or proposals to increase Iranian transparency will discourage its attempts to acquire more potent technologies including those involving outer space. Iran has circumvented sanctions in the past. As has been observed, most States resent restraints imposed on their attempts to acquire external influence.

Israel

Jewish religion and culture has a long history. But it is difficult to determine the influence of religion on modern Israel, given the additional impact on Jewish attitudes of more contemporary events. These include the German-initiated slaughter of Jews and other vulnerable minorities in the 1930s and 40s, together with the creation of the modern Israeli State from the Mandate of Palestine and the conflicts that followed.

Modern Israel is a geographically small State (approximately 21 000 km^2) surrounded on one side by ocean and on three by States hostile to it. Israel repelled attacks, either initiated or threatened against it, in 1948, 1956, 1967, 1969, 1973 and 1982 in addition to confronting many less serious and spasmodic conflicts with the residual Palestine. Israel's most significant diplomatic, military, and economic relationship is extra-regional and is with the US.

Israel tries to balance an apparent desire for independence and self-determination with the reality that the US provides support through both arms and financial assistance. Israel's attempt to promote its nascent military and strategic independence through indigenous arms production and sales has therefore led to occasional tensions with the US; for example when Israel has exported some technologies to the PRC.[54]

The relationship between the Israel Space Agency and Israel's other government and private sector agencies is unclear. The Agency was formed in 1982. In 1984 the then Israeli Minister for Defence, Moshe Arens, ordered the start of a domestic launch vehicle and satellite construction program involving Israel Aircraft Industries, now renamed Israel Aerospace Industries, and several other companies. The Space Agency presumably had a policy formulation role. It subsequently may have engaged with the national space program through coordination either in an advisory or executive capacity.

Israeli industry and government agencies now produce a range of Earth observation satellites and can also build indigenous launch vehicles. These initiatives have involved cooperative arrangements with France, US, Canada, India, Germany, Ukraine, Russia and The Netherlands. In 2011 Israel signed a cooperative agreement with the European Space Agency (ESA).

Despite the Israel Space Agency's website disclosing nothing about military operations it seems clear from data and information already considered that Israel has in place a policy and national security program identifying outer space as a potential battleground. In 2005, Yuval Steinitz, the chair of Israel's Defence and Foreign Affairs Committee was recorded as calling for the development of specific anti-satellite weapons.[55] Cumulative evidence suggests Israel might be able to launch a nuclear weapon into low Earth orbit or will be able to do so in the foreseeable future. It either has access to or could develop an anti-satellite kinetic weapon. Given its experience with space launch, and with remote control technologies such as are employed with UAVs, Israel has mastered key technologies needed to build and operate a space plane.

Regardless of Israel's outer space-related policies and ambitions, its capacity to independently prosecute warfare in the space region is and will almost certainly remain geographically constrained. If an Israeli launch vehicle adopted an east, north or southern inclination it would overfly hostile territory, conceivably attracting retaliation, particularly if the launch occurred during a time of heightened tension. Launching to the west, across the Mediterranean Sea, requires launching against the direction of Earth's rotation which, among other disadvantages, places into orbit spacecraft travelling in the opposite direction to other satellite traffic and related debris.

Westward launches also constrain space warfare options by making co-orbiting vehicles almost impossible and increasing fundamentally all target acquisition challenges, given the increased velocities and closing speeds relative to many other spacecraft. An alternative for Israel would be to persuade compatible States with launch facilities to actively support Israel's military ambitions, although reliance on this tactic seems imprudent.

Israel has adopted a policy of opacity in relation to its nuclear weapons. This attitude appears to have included other military capabilities, and particularly those such as missiles with implications for space warfare. However, Israel's location in a geopolitically unstable region and its proven determination and capacity to defend itself against extant or perceived threats, thus far successfully, all point to a potential for Israel to take aggressive action in the outer space region if it perceives advantages from doing so. That situation would most likely arise in response to a missile attack from or through outer space, in which case almost vertical interception launches from Israel would reduce the likelihood of triggering potentially aggressive regional responses.

Japan

Building upon a strong and ancient intellectual and cultural legacy, Japan began modernising in the 19th century, arguably the process beginning with the 1868 Meiji restoration. But modernisation failed to dilute a traditional military disposition and by 1894 Japan was at war with the Chinese empire, followed by war with Russia in 1905. In addition to numerous skirmishes Japan also annexed and occupied Korea in 1910, fought against Germany in the First World War, invaded the Manchurian region in 1931 and other areas of China in 1937, before attacking the US in 1941. Japan's war against the US and other Asian and Pacific States ended with Japan surrendering in 1945 following the US nuclear bombings of Hiroshima and Nagasaki. Between 1945 and 1952 Japan was occupied by an international force led by the US.

Japan entered the second half of the 20th century as a militarily defeated nation distrusted throughout Asia but soon to become allied closely with the US. The US, recognising an opportunity to contain the USSR and later the PRC, established military bases on Japanese territory, Okinawa being the most populous.

Following nuclear bombings of the Second World War Japan campaigned diplomatically for limitations on nuclear weapons, although official enthusiasm within Japan for maintaining that stance may recently have weakened.[56] Nuclear weapons aside, Japan has been creating a potent military capacity based on its 237 000 military personnel armed with modern equipment and supplemented by 38 000 US military personnel. Japanese defence expenditures appear small (typically less than 1.0 per cent of GDP) but the Japanese GDP is large (2013 estimate USD 4.729 trillion). Japan has thus been able to acquire and develop a range of high technologies linked to the contemporary battlefield.

Japan's space policy at first reflected a strong commitment to utilising outer space for peaceful purposes. One early indication of a changing approach occurred during the year 2000 when the Council of Science and Technology Policy realigned strategic space planning in Japan by moving it from the Ministry of Education, Culture, Sports, Science and Technology into a community linked to the Prime Minister's office. In 2008 the apparent nexus between pacifism and outer space activities was modified by Japan's *Fundamental Act of Outer Space* (2008) which allowed limited use of outer space for security-related purposes. Article 24 of the Act required an Outer Space Exploitation Strategy Headquarters to produce an Outer Space Master Plan. I did not locate the plan, which might have been classified for security purposes; instead I uncovered a *Basic Plan for Space Policy* (2009).

The Basic Plan included the objective of contributing to the enhancement of security although elaborating comments were confined to '*strengthening information gathering*' supplemented by a reference to '*maintaining our exclusively defense-oriented policy*'.[57] On 14 February 2012 the Japanese government was reported as having established a space policy panel in its Cabinet Office.

Japan has been developing missiles for more than 70 years, starting with the Funryu series during the Second World War. At that time the Imperial Army also embarked on the I-go Project, involving a radio-controlled missile designed for release from beneath the wing of an aircraft.

Work with missiles continued post-war. First was construction in 1955 of the Pencil, a 23 - 30 cm, 200 g micro-rocket. Then followed the Baby (105 -120 cm and 10 kg), and after that the Kappa, Lambda, Mu, Q, N and other increasingly larger and more powerful vehicles in an evolutionary sequence. One version of the current H II launch vehicle, the *H2A204*, with a length of 53 m and a mass of 443 tonnes, is capable of lifting six tonnes to GEO transfer orbit.[58]

The development of capable space-launch vehicles left no doubt Japan possessed intercontinental ballistic missile capabilities including the capacity to launch nuclear weapons into outer space should Japan develop or acquire them. Japan has also acquired at least six destroyer-class ships equipped with the US Aegis radar system which supports SM-3 missiles. An SM-3 missile fired from a Japanese ship destroyed a targeted missile '*above the atmosphere*' in October 2009, the interception being aided by tracking support from a US vessel.[59] Other Japanese developments directly relevant to space warfare were described in previous chapters and included advanced co-orbital capabilities and docking technologies. Japan clearly has capacity to intercept and attack spacecraft.

Japan's transition from a pacifist policy to one in which defence and national security are more prominent might be a two-speed process. Technology, particularly technology supporting space warfare capacity, may have advanced ahead of published national policy. A significant but unknown factor in relation to hostilities in space is whether Japan's security relationship with the US might draw Japan into a conflict as an extension of alliance arrangements. Also unclear is whether Japan would be permitted to operate its SM-3 missiles in the anti-spacecraft role without US approval or support.

Pakistan

The inclusion of Pakistan in this cohort of spacefaring States was justified more by its potential for space engagement than on its modest achievements to date.

The 8th century CE traders introduced the Muslim religion to people occupying the geographic region later known as Pakistan, and Islamic traditions have persisted. Religious beliefs appear to have justified formation of the Muslim League movement in 1906 with the primary purpose of ensuring India's Hindu majority could not dominate the Muslim minority.

Pakistan, both East and West, became independent from Britain in 1947, subsequently dividing again in 1971 with the creation of Bangladesh in the eastern region.

Under the 1947 independence arrangements Britain had agreed that regional princely states could join either India or Pakistan. The Maharaja of Kashmir equivocated, eventually signing accession papers in favour of India. Pakistan refused, and continues to refuse, to recognise that accession. Pakistan's subsequent history of antagonism toward India has been influenced by this disagreement but it has also been in part fuelled by continuing religious differences, economic challenges, political instability commonly involving the Pakistani Army and inconsistent relations with the US as well as by Pakistan's positive relationship with the PRC.

In 1954, concerned over India's economic and military dominance, Pakistan entered with the US a *Mutual Defence Assistance Agreement*. A further bilateral arrangement, the *Pakistan-US Agreement on Cooperation*, was negotiated in 1959, allegedly assuring Pakistan of US support in the event of aggression. However Pakistan was disappointed when it attempted to invoke the Agreement in 1965. The US responded by resisting involvement on the grounds that events in question did not oblige it to intervene.[60]

Subsequently, Pakistan's relationship with the US deteriorated in the mid 1970s following US opposition to Pakistan's nuclear program, improved when Pakistan opposed the USSR's invasion of Afghanistan, deteriorated again in 1997 when Pakistan recognised the Taliban government of Afghanistan, then improved yet again with Pakistan's support for the US-led 2001 intervention in Afghanistan.

The Pakistan-India relationship has remained turbulent. There was armed conflict between them in 1971. It was followed by an alleged Indian plan to attack Kahuta in 1984, the Brasstacks Crisis of 1986 - 7, the Kashmir Crisis of 1990, India's renewed nuclear tests in 1998 and other intermittent events. India and Pakistan have both exhibited satisfaction and patriotic pride when declaring their nuclear weapons capabilities. Pakistan nevertheless entered the *Agreement between India and Pakistan on the Prohibition of Attack Against Nuclear Installations and Facilities* (India -Pakistan Non-Attack Agreement (1991)) in which each nation agreed to not attack each other's significant infrastructure projects or nuclear facilities.

Pakistani space programs are coordinated by the Pakistan Space and Upper Atmosphere Research Commission which serves as the national space agency of Pakistan.[61] So far as I could determine Pakistan's engagement with outer space has reflected a pragmatic assessment of requirements for improved land management and agricultural production together with research to support these objectives. In February 2012 the then Prime Minister of Pakistan, Syed Yousuf Gilani, approved a Pakistan Space Policy to be presented to the National Assembly. As at early 2014 the policy still could not be found on publically available websites.

Pakistan has never launched a satellite from an indigenous launch vehicle.

Evidence presented earlier indicated Pakistan's actual or potential capacity to launch a nuclear or any other weapon into space was at best marginal but could not be dismissed unconditionally. Any such attempt might require support and conceivably permission from the PRC or DPRK. No data or other information was found to suggest Pakistan was consciously developing a capacity to engage in space warfare.

USSR-Russia

Russia formed as a recognisable State during the 9th to 11th centuries CE although a Mongol invasion in the 13th century delayed consolidation. By the 17th century, Russia had become an imperialist entity which began increasing its influence eventually reaching east to the Pacific Ocean, west to incorporate components of Europe, south to include parts of Asia, and north to the ocean shores.

An unsuccessful war with Japan together with increasing social unrest in the early 20th century preceded the revolution of 1917. It was fuelled by communist idealists and led to the formation of the USSR. Antagonism between the USSR and the West began almost immediately with the US, France, Britain and Japan sending a force of more than 100 000 troops to overthrow the new government. The attempt failed.[62]

Following the revolution of 1917, USSR foreign relations practitioners faced a dilemma. States openly antagonistic to the communist government needed to be contained or discouraged and this demanded limited security arrangements with, for example, Germany (the *Treaty of Rapallo*). However Marxist-Leninist philosophy, the motivating force behind the USSR's political doctrine, opposed the very notion of States thus limiting opportunities and motivations to establish a diplomatic network.

The USSR's communist rule was subsequently strengthened and industrialisation implemented at the cost of millions of its citizens' lives. During the 1930s, as war in Europe appeared imminent the USSR Premier, Iosif Vissarionovich Dzhugashvili (known in the West as Joseph Stalin) simultaneously began negotiating an anti-German alliance with France and a non-aggression pact with Germany. An agreement with Germany, the *Molotov-Ribbentrop Pact* of 23 August 1939, was eventually signed but on 22 June 1941 Germany nevertheless invaded the USSR.

During the Second World War, political pragmatism overwhelmed communist philosophy. Cooperative logistics, military and broader strategic arrangements were agreed between the USSR, Britain and the US, notably at the Tehran conference of November 1943 and the Yalta conference of February 1945. However, significant separation was maintained between the USSR and other allies. The Eastern and Western command structures were never integrated. Following the Second World War, cordial relations between the USSR and the US dissolved and were replaced by an increasingly hostile confrontation.

By the end of the Cold War in 1991 internal attempts to restructure and modernise the communist empire had failed. The USSR fragmented into component parts, essentially Russia and 14 republics. Antagonism between Russia and the US then eased but mutual distrust remained at variable and occasionally significant levels. Meanwhile, Russia had established a comprehensive global diplomatic network.

Publicity regarding the US achievement, in 1969, of successfully landing the first people on the surface of the Moon and returning them safely to Earth, might have overshadowed USSR technological achievements in the outer space region. The USSR was the first State to launch an artificial satellite into orbit, first to launch a living animal into space, first to land a vehicle onto the Moon, first to record the far side of the Moon, first to orbit a human in space, first to undertake a human spacewalk, first to make a soft landing on the Moon, first to send a spacecraft into the atmosphere of Venus, first to land on Venus, first to autonomously land on the Moon and retrieve geological samples for return to Earth, and first to orbit a space station. The USSR/Russia has consistently built high-quality engines to power space-launch vehicles, some having been imported by the US.

The Russian document, *The Foreign Policy Concept of the Russian Federation* (2008), places national security uppermost in strategic priorities, specifically mentioning territorial integrity and an ambition to achieve authority in the world community. Russian policy is to cooperate internationally and constructively with States in all regions and on all continents across a range of initiatives. The Foreign Policy Concept document refers to the US and PRC in generally positive terms.

An additional document, *National Security Concept of the Russian Federation*, is less subtle. It refers for example, to the threat of '... *attempts to create an international relations structure based on domination by developed Western countries in the international community, under US leadership, and designed for unilateral solutions* ... '[63] Threats identified in the international sphere included NATO's eastward expansion and NATO's shift to the practice of using military force outside its zone of original responsibility, without UN Security Council authorisation.

Russia's space agency, Roscosmos, has published its *Status and Functions* which appear to represent Russia's national space policy.[64] The document portrays essentially peace-oriented objectives with emphasis on administrative and coordination responsibilities. It includes reference to, '*Jointly with the Defense Ministry [ensuring] launches of spacecraft for military purposes*.' But no elaboration of that concept is given. One pragmatic aspect of Russia's response to a perception of Western expansion was establishment of a Space Defence Troop, a new combat arm incorporating the Russian space and missile defence commands, although further reorganisations may have followed.

From data and information included previously it is apparent Russia has inherited from its USSR predecessor a stock of nuclear weapons and the launch vehicles able to deliver then into or through outer space. The USSR also had experience with detonating nuclear weapons in outer space.

Russia additionally has anti-satellite kinetic weapons and advanced co-orbital capabilities, including the capacity to attack large satellites, as demonstrated when it de-orbited its own space station using the Progress device. USSR experience with its space shuttle, the Buran, must have equipped Russia to begin building a viable space plane. Some evidence supports the possibility that Russia is developing hypervelocity vehicles.

Low-powered lasers aside, no evidence was located to indicate the USSR, or Russia, had ever developed a destructive beam weapon for use in outer space although it apparently attempted to do so. It has had experience with orbiting space-based nuclear reactors, which might be developed to power such weapons if the weapons ever become practicable.

Russia's national history, relations with the West, its policies and its space warfare technologies all follow a comprehensible sequence. Russia has both the motivation and the technological competence to maintain an influential space warfare capacity into the foreseeable future. In February 2012 Russia announced a 10-year program to provide over 400 modern ICBMs and more than 100 military spacecraft.

Europe

Although not a nation State, the dynamic and loosely-defined region *Europe* presents a unique and complicated situation in relation to warfare, including the potential for warfare in outer space.

The *North Atlantic Treaty* (1949) included agreements to maintain and develop individual and collective capacity to resist armed attack and to regard an armed attack against one or more NATO members, in Europe or in North America, to be considered an attack against all of its members. Article 6 of the Treaty purported to limit the concept of *attack* to events affecting territories, forces, vessels or aircraft of any of the Parties to the Treaty. However, the space-related Liability Convention (1972) assigned continuing State responsibility for '*space objects*'. It therefore seems likely that NATO member States could regard an attack against a space object registered to any one of them to fall within the terms of their Treaty, if they elected to do so. But, some European States are members of NATO; others are not.

Additionally, some European States are members of the European Union (EU) which in its present form emerged from the *Treaty of Maastricht*. That Treaty has evolved to extend beyond economic arrangements and to take account of cooperation on foreign policy and security. There can be no certainty that the interests of NATO members will invariably coincide with those of some or all of the other EU members. The potential for complications has been further increased by some States adopting interrelated monetary policies. A common currency, the Euro, created an additional bond via the Eurozone.

The European Space Agency (ESA) was formalised in 1980 with a membership that did not (and still does not) mirror that of NATO, the EU or the Eurozone. ESA is guided by the *ESA Convention* which professes exclusively peaceful purposes. Member States of ESA, France in particular, may nevertheless have evolved their own space warfare capabilities as previously considered. The ESA Convention does not require member States to surrender any unilateral national security

The following table, Table 7 – 2. *Member States of NATO, the EU, the Eurozone and ESA* points to complicated Europe-centric strategic and international relations, given the diverse memberships of the separate but sometimes interrelated organisations already mentioned. Of the 35 States listed only nine are members of all four organisations shown. Whether those States could provide a core of stability in case of dissent regarding military operations is unknown.

Table 7 – 2. Member States of NATO, the EU, the Eurozone and ESA. (Compiled **February 2014**) [65]

	NATO Member State (Y/N)	Member of European Union (Y/N)	Within Eurozone (Y/N)	Member of European Space Agency (Y/N)
Albania	Y	N	N	N
Austria	N	Y	Y	Y
Belgium	Y	Y	Y	Y
Bulgaria	Y	Y	N	N
Canada	Y	N	N	Cooperating
Croatia	Y	Y	N	N
Cyprus	N	Y	Y	N

	NATO Member State (Y/N)	Member of European Union (Y/N)	Within Eurozone (Y/N)	Member of European Space Agency (Y/N)
Czech Republic	Y	Y	N	Y
Denmark	Y	Y	N	Y
Estonia	Y	Y	Y	Cooperating
Finland	N	Y	Y	Y
France	Y	Y	Y	Y
Germany	Y	Y	Y	Y
Greece	Y	Y	Y	Y
Hungary	Y	Y	N	Cooperating
Iceland	Y	N	N	N
Italy	Y	Y	Y	Y
Ireland	N	Y	Y	Y
Latvia	Y	Y	Y	Cooperating
Lithuania	Y	Y	N	N
Luxembourg	Y	Y	Y	Y
Malta	N	Y	Y	N
Netherlands	Y	Y	Y	Y
Norway	Y	N	N	Y
Poland	Y	Y	N	Y
Portugal	Y	Y	Y	Y
Romania	Y	Y	N	Y
Slovakia	Y	Y	Y	N
Slovenia	Y	Y	Y	Cooperating
Spain	Y	Y	Y	Y
Sweden	N	Y	N	Y
Switzerland	N	N	N	Y
Turkey	Y	N	N	N
Britain	Y	Y	N	Y
United States	Y	N	N	N

In relation to military activity affecting the outer space region, complications are already emerging for Europe and several component States. The US is a member of NATO and can be expected to influence NATO members in relation to engagement in warfare generally, including any prospective warfare in outer space.

One example of this influence has been US proposals to install a missile defence shield in Europe, apparently under the auspices of NATO.[66] If the missiles installed now or through improved models in the future could be adapted for an anti-satellite capacity then some European States formerly unable to engage satellites may be enabled to do so.[67] This seems possible, even likely. Under the US '*phased adaptive approach*' to the installation of missiles in Europe, installations will include the SM-3 missile a variant of which has been used to intercept and destroy satellites.

While NATO binds some European States to the US and *vice versa*, those European States which are members of NATO as well as members of the EU confront at least one significant and growing dilemma. The EU is seeking increased independence from the US, a trend given practical effect through acquiring local global navigation satellite services (GNSS). The EU and ESA are developing the *Galileo* satellite constellation with both entities contributing to its funding. In 2003 the EU signed an agreement with the PRC, potentially the US's main competitor in outer space, allowing the PRC to become a member of the *Galileo* joint undertaking as from October 2004.

In contrast with the ESA Convention's exclusively peaceful purposes clause, guidelines for an evolving European space policy include a need to meet Europe's space-based security and defence needs.[68] The extent to which member States might subordinate national interests to this common cause of European defence and security, the conditions under which they might do so and the implications for European States which are also NATO members, are all unknown.

Many of the conclusions I might draw from this matrix of conflicting issues would not be evidence-based. Among the issues not apparently resolved are: the mature-state capabilities of European missile defence systems, whether participating States will have capacity to operate missiles without NATO approval and whether NATO approval will be critically dependent upon US endorsement. A requirement for US endorsement could be overt, for example through a formally negotiated arrangement, or covert if missile operation required access to software codes held only by the US.

It seems possible, perhaps probable, that influences exerted by the existing matrix of multilateral security and diplomatic arrangements have reduced the capacity of any

European State to unilaterally engage in space-related warfare, although the extent to which this has occurred is not known.

Summary

Identifying the space policies of States is a challenging and imprecise process, particularly as no coherent and widely accepted concept of an outer space policy has been articulated, a feat not achieved by this work either. The foreign relations activities of States were examined through a prism which separated their observable activities into three primary but connected spheres: expansion of influence, maximisation of influence and responses to impediments. The implicit conclusion was that States engaged in advancing their foreign policy objectives were commonly motivated by a desire to expand their power and influence internationally, and in a competitive environment, to the extent possible given prevailing circumstances. Those policy-related behaviours, singly or in combination, set the stage, in part, for conflict in outer space.

The list of single-State commitments to international agreements that were components of international law-related warfare, (Table 7 – 1. *Selected International Agreements; Agreements, Signatures, Ratifications, Accessions and Successions and Non-Party Status*) revealed little. No significant conclusions could be safely drawn from the data.

Consideration of the strategic histories and space policies of specific States was more illuminating, particularly for those States exhibiting a strong association between historical or cultural influences and contemporary behaviours, with possible implications for likely future conduct.

Specifically:

a. The US is accustomed to achieving results substantially on its own terms and is unlikely to change that expectation. The US has adopted a defensive and offensive posture regarding potential conflict in outer space. Its policies are open and candid. They are compatible with its military record. The potential for the US to become involved in a conflict in outer space may be higher than that of most other States and the US appears to be providing for this contingency.

b. Brazil benefits from a non-aligned posture which requires positive relations with many other States. Brazil has had a variable technological capacity to participate in outer space conflict. By 2014 progress toward acquiring this capability seemed to have stagnated. Brazil is most likely to pursue its declared interest in commercial space enterprises while avoiding conflict in the outer space region.

c. Britain, lacking an indigenous space-launch vehicle, has become and is likely to remain an insignificant participant in space warfare except as an ally of the US.

d. The political and strategic experiences of the PRC, and of China beforehand, have provided few incentives for it to hold the West in high regard. As its power and prestige increase the PRC may become increasingly competitive and assertive. The PRC's apparent space policies, although less candid than those of the US, allow for the development of defensive and offensive space capabilities. The PRC is a potential participant in outer space conflict.
e. The DPRK is an enigmatic State. A rational approach to regional stability and economic welfare would have it building cooperative relationships with Japan and RoK, enhancing its economy, abandoning nuclear weapons and relying on the PRC for support in case of a missile attack. It is doing none of those things except through token initiatives. The DPRK's unfathomable attitudes combined with an ambition to develop long-range missiles or space-launch vehicles categorise the State as a potentially dangerous and destabilising agent in global space affairs.
f. As an expanding regional influence, India could encounter opposition or competition and it has, conceivably in anticipation, established large military forces. While officially India remains committed to the development of commercial space assets, sufficient information is available to indicate it is also attempting to build an offensive or defensive space-weapon capacity with implications for potential future engagement. However, aside conceivably from the PRC, potential space-related allies and opponents are not clearly visible.
g. Iran is confronting a challenging situation, having either alienated or at least alerted States in western Europe and the Middle East as well as the US to its potential for engaging in military action. Iran has not revealed a comprehensive strategic space policy. Confusing and conflicting indicators suggest Iran is a State to be monitored but not one necessarily likely to act rashly with regard to security of the outer space region.
h. Israel pragmatically comprehends its geopolitical situation and understands its national survival has been and will remain threatened. Its known launch vehicles, technological skills and military prowess leave little doubt that Israel would take a conflict into outer space to confront a serious threat. One of several concomitant challenges for Israel is that any conflict threatening its survival is likely to be regional, but warfare in outer space could rapidly become global in its implications and consequences.
i. Traditionally militaristic, Japan has started to confront the aberration of a post-war passive defence policy. Its conservatively-worded space security policy appears to lag behind technological developments which already have armed Japan with a potent capacity to conduct warfare in outer space. How and against whom such a capability might be directed is unclear.
j. Pakistan faces almost insurmountable strategic challenges. Surrounded by a combination of more powerful States, PRC and India in particular, and

periodically unstable States, notably Afghanistan and occasionally Iran, Pakistan is additionally burdened by a depressed economy. Whether its most potent challenges will arise from domestic or external influences is difficult to know. Pakistan's space and defence programs could in the future provide a basic space weapons capability but Pakistan is unlikely to be able to deploy it without risking devastating retaliation.

k. The USSR, subsequently Russia, was the foremost space pioneer. Stemming from that experience and spurred by the Cold War, Russia acquired a suite of effective space weapons. Whether Russian military assets, including space assets will eventually be deployed independently from, or in an alliance with, or against, the PRC cannot yet be determined.

l. European States have become strategically fragmented by the superimposed and not fully compatible structures of NATO, the EU and ESA, influenced to some degree by the Euro Zone. A fundamental technical competence to engage in outer space hostilities is resident within the continent. But the legal, diplomatic, economic and administrative maze which many European security-related proposals must navigate might not permit effective corporate participation.

States most likely to become directly involved in a conflict fought in outer space are the US, PRC and Russia. Japan or Britain might be drawn into such a conflict as a consequence of their security arrangements with the US. India and Brazil are likely to remain non-aligned and less likely to engage in space warfare. The potential for Israel, Iran or Pakistan to engage in space hostilities cannot be dismissed unconditionally. The DPRK could act irrationally and with minimal warning.

This dialogue provides a geopolitical component for the following chapter which is focussed predominantly upon issues involving the US, PRC and Russia on the one hand, and the majority of almost voiceless middle-level and smaller States affected by activities in outer space on the other.

Chapter 8. Containment Strategies and Tactics

Previous chapters traced elements of the recorded history of human warfare and: analysed how evolving capacities to wage war had eroded distinctions between conflicts involving the land, sea, air, sub-surface and outer space; described categories of outer-space weapons; considered the influence of international law on outer space; and examined States' outer space policies against key elements of their historical and geopolitical backgrounds. Cumulatively, that dialogue supported the notion that humanity, primarily through the machinations of nation States, is accustomed to violence or the threat of using it as a conventional and often convenient foreign policy technique. The evidence thus pointed to a likelihood that potential for warfare in outer space will remain unabated unless attempts are made to discourage and contain it.

Regardless of the negativity inherent in that analysis, the facts remain that outer space is a region from which many valuable services to humanity can be and are now being delivered, a region in which no war has yet been fought despite the established capacity to do so, and a region vulnerable to pollution with potential to deny or constrain human access for periods lasting from weeks to centuries. So there exist both incentives and opportunities to reduce or manage associated tensions.

The PRC, US and Russia are considered to be the major spacefaring States as well as those most likely to participate in space warfare although other States also have capacity to enter any such conflict, albeit on a smaller scale. Those three have a record of engagement with most or all forms of outer space-related activity, extending from diplomatic initiatives to development of a range of weapons able to be deployed into outer space. They also share competitive ambitions to influence the outer space region.

Adding to the potential for tensions to arise, there are obvious if imprecise limits on the extent to which those major spacefaring States can constructively affect each other's conduct. Neither Russia nor the PRC is likely to voluntarily endorse US leadership in space. It is identically inconceivable the US would contemplate ceding that same concession to any competitor State, as has been advocated by the PRC's Ministry of Foreign Affairs for example.[1]

In this context, every anthropogenic object in Earth-centric orbit is a representation of technologies, policies, insecurities and ambitions originating within a terrestrial framework, commonly from within an environment fostered by one of the major three States. The link between human-initiated events on Earth and their physical expressions in outer space therefore raises the spectre of entwined diplomatic, military, economic and political perceptions combining to produce insular and State-centric perceptions of space power.

Those perceptions have evolved despite there being no precedents to indicate the likely progress of a significant conflict in outer space. The effectiveness of some categories of space weapon is neither known nor knowable. A major space war involving weapons demonstrably able to damage or destroy spacecraft could result in the one or all of the major States, losing both power and prestige. Several other States also have capacities to render the outer space region inaccessible to every nation, or less accessible than at present, for protracted periods. Space warfare thus confronts even more uncertainties than those revealed in Chapter 1 as affecting terrestrial contests, uncertainties amplified by irrefutable records of overtly weaker and more vulnerable combatants having sometimes emerged victorious.

These issues contribute to the insecure foundations upon which the military and related policies and activities of spacefaring nations now rest.

Preserve Freedom

The most prominent three spacefaring States continue to exert influence over the outer space region, as well as in other environments, through a spectrum of initiatives including the activities and strategies I will now consider.

Foerster, 2012, coined the term *entanglement* to describe the web of agreements and commitments negotiated by prominent States to reduce the potential for opposition from each other or from some of the less powerful States.[2] However, the concept of entanglement applies to States other than direct competitors. States with broadly shared strategic values are commonly described as *allies* but they are nevertheless convincingly entangled. Examples of such arrangements and proposed arrangements are numerous. They have included *The North Atlantic Treaty*, the *Security Treaty between Australia, New Zealand and the United States* (ANZUS), Russia's proposed *European Security Treaty*, Russia's *Collective Treaty Security Organisation* and the *DPRK-China Treaty of Friendship, Cooperation and Mutual Assistance*. Some of these arrangements can be interpreted as extending to space-related activities. Singly and in combination they represent opportunities for more prominent powers to gain congregations of client States whose operational sovereignty becomes fundamentally diluted.

Benefits to major States can be particularly significant where international relationships include military cooperation. At a minimum, client States are committed to not endorsing the military initiatives of perceived competitor States. On occasions clients will lend additional support through, for example, compatible voting in the United Nations and other forums or by providing military forces to bolster powerful State ambitions. Trusted allies can be invited to host intelligence-gathering installations thus integrating the client States into a shared strategic vision based in part on common data and information.

Arms sales to client States provide further entanglement opportunities. The sales exacerbate dependencies by client States for technological support, replenishment of munitions and access to upgraded software suites or technologies as they become available. The consequent framework offers provider States both financial benefits and political influence. Pointing to a growing trend, between 2004 and 2011 arms sales to developing States increased at a rate greater than did sales to developed States.[3]

Entanglement initiatives are often represented as the exercise of diplomatic or economic power, but diplomatic power can change quickly into dominant influence when client States are penalised for not conforming to major-State policies. In 1985 a small Pacific nation, New Zealand, exercised its sovereign right to insist visiting ships not be nuclear powered or carry nuclear weapons. The US reacted by severing military and intelligence ties with New Zealand, despite New Zealand having provided military support to predominantly US forces in the Korean and Vietnam Wars, and regardless of the ANZUS Treaty being in force.[4] However, retribution is not an inevitable outcome. Canada avoided significant punishment when it provided residency for Americans avoiding compulsory US military service.

Not all entanglement arrangements evolve as originally intended. Russia signed a cooperative agreement with Germany, the *Molotov-Ribbentrop Pact* of 23 August 1939, but on 22 June 1941 Germany invaded the USSR. In 1972 the US and USSR entered into the *Treaty between the United States of America and the Union of Soviet Socialist Republics on the Limitation of Anti-Ballistic Missile Systems.* On 13 December 2001 the US advised Russia of its intention to unilaterally withdraw from the treaty. A complicated network of agreements can also lead to conflicting obligations, as appears to have happened when the US was reluctant to confront Pakistan regarding its India-alleged role in at least two terrorist-related incidents.

Despite uncertainties, new and revised agreements are under almost constant negotiation. Between 2005 and 2010 the US arranged three security-related agreements with India. Russia has been working since 1995 to negotiate a new agreement for security arrangements with the European Union. Other initiatives are being progressed.

Regarding multinational arrangements for arms control in outer space, US policy is that it will consider proposals for arms control measures if they are equitable, effectively verifiable, and enhance the national security of the United States and its allies.[5] No past or present multinational treaty or agreement relating to outer space could have met all those prerequisites, in particular the verifiability requirement, and so US policy is tantamount to a refusal to negotiate unless the outcome will enhance US power or influence. The PRC has adopted a similar but possibly more flexible approach, its 2011 White Paper on space recording the objective of, '*... protect China's rights and interests, and build up its national comprehensive strength*.'[6] Similarly, Russian space policy includes the objective of, '*... ensuring defence capabilities of the Russian Federation and control over the implementation of international treaties concerning armaments and armed forces*.'[7] In all three examples the common denominator is national interest, not global stability within a global environment.

Minimisation of constraints to activities in outer space has been one of the factors enabling competitive development of space weapons and other disruptive technologies designed for deployment into outer space. Whether prohibition by international law would have prevented such activity is doubtful; however, it could have enabled States conducting activities categorised as illegal to be identified and chastised thus applying the subtle but pervasive pressure available to the international community through international law.

The major spacefaring States claim, or imply, they already have a credible capacity to respond in outer space to attacks occurring there. The *US National Security Space Strategy* goes further by recording circumstances under which the US will implement its retaliation capacities.[8] The PRC and Russia have, typically, been more opaque regarding retaliation plans.

The concept of a credible response to an attack against an orbiting spacecraft assumes a demonstrated capability to act, combined with a reasonable likelihood it will be used if necessary. From previous chapters it is apparent the major spacefaring States have inventories of weapons, notably nuclear, kinetic and co-orbital devices, enabling physical responses to attacks against their space vehicles. Some other States have been developing similar capabilities.

But notwithstanding the availability of these and other space weapons, responding credibly in outer space will be challenging, particularly so for any State hoping to reduce the prospects of uncontrollable escalation. An aggrieved State will need to determine that an intentional attack against its asset in outer space did occur or, more controversially, was imminent. In the case of an alleged attack, there may a requirement to quantify claimed damage before identifying an appropriately proportional response. It will be necessary to either identify the attacker with verifiable confidence or risk retaliation if a mistake is made or even claimed by an opponent to have been made. The notion of a capacity to retaliate, credibly or otherwise, is complicated.

Additionally, nuclear or kinetic weapons if used in outer space will likely result in debris clouds and potentially attract opposition from States owning assets affected collaterally by debris, by NDRE or both.

The combination of these factors implies that in some, or possibly most, circumstances terrestrial retaliation might be a preferred response to one risking an escalation in outer space. Conversely, if strategically significant spacecraft were destroyed, apparently in a deliberate attack, responses intended to deny a competitor access to a similar data stream or service might be judged appropriate. Terrestrial responses cannot therefore be assumed.

Emphasising the conceivably misleading notion of credible retaliation, space warfare-related strategies, doctrines and tactics are periodically rehearsed. These activities imply that the threat is realistic and conveniently add political weight to claims, mainly by national security and defence constituencies, that ways to counter it should be resourced. Related development and testing processes also help promulgate, domestically, the impression that a State has available a viable retaliation plan.

Comprehensive details of rehearsal or simulation arrangements relating to conflict in outer space are not revealed publically, adding further to the mystique and associated impression of significant capability. Some information has nevertheless become accessible on broader strategic intentions with less available on tactical plans and almost no data released on specifics such as the military rules-of-engagement in outer space.

The US, typically, has made available more information on its strategy, doctrine and tactics than has any other State. Concepts fundamental to US space strategy are listed in *National Space Policy of the United States of America.* US space warfare doctrines have also been published. Potential situations and military responses to them are rehearsed in war-gaming exercises which have been extended to include participation by representatives of States allied to the US.[9] Russia inherited from the Cold War era a comprehensive space strategy and doctrine which remains under development as illustrated by its formation of a specific space force in 2011. China has been less transparent, but it is obvious from literature available that strategies and doctrine relating to warfare in outer space have been evolving over decades.[10]

It is therefore appropriate to conclude that the PRC, US and Russia, as well as possibly other States believe they have prepared, doctrinally and organisationally to conduct and defend against combat operations in the outer space region. However, for reasons already highlighted the precise nature of any potential military confrontation in outer space is at present unclear. This uncertainty casts doubt on the validity and utility of strategies, doctrines and tactics that are rehearsed.

Nevertheless, related activities are continuing and the concept of credible retaliation capacity is likely to be increasingly associated with two developments occurring within the major spacefaring States; improving electronic and related attack technologies and developing 'space situational awareness', both of which are now reviewed.

From information and data already provided it appears, despite obstacles, the US, the PRC and Russia continue to invest in developing, or attempting to develop, weapons to attack satellites by using electromagnetic and associated techniques. Electronic attacks should commonly offer two benefits when compared against attacks involving nuclear or kinetic weapons. Unless the target has been salvage fused, an electronic attack will result in minimal debris. Additionally it can be difficult to detect and locate the source of an electromagnetic-energy or related attack applied over a possibly brief period of time. These circumstances may create opportunities for unattributable or less-attributable attacks.

But all potential combatants are likely to have similar capabilities. Moreover, spacecraft are being designed to resist harm or interference from radiofrequency and related sources in a cat-and-mouse game that has persisted since long before the space age began. Success when using these techniques to attack spacecraft cannot therefore be assumed. Consequently, the major spacefaring States, and some other States, will almost certainly retain the capacity to deploy nuclear, kinetic or related weapons in outer space. Ostensibly these may be developed and represented as components of missile defence, but they obviously have broader potential applications for outer space conflict.

The major spacefaring States, supported cooperatively by entangled or client States, have also installed 'space situational awareness' (SSA) facilities to collect data from outer space on a spectrum of phenomena. These are shown in the following diagram.

Figure 8 – 1. Primary Sources of Observations Contributing to a Space Situational Awareness Architecture. [11]

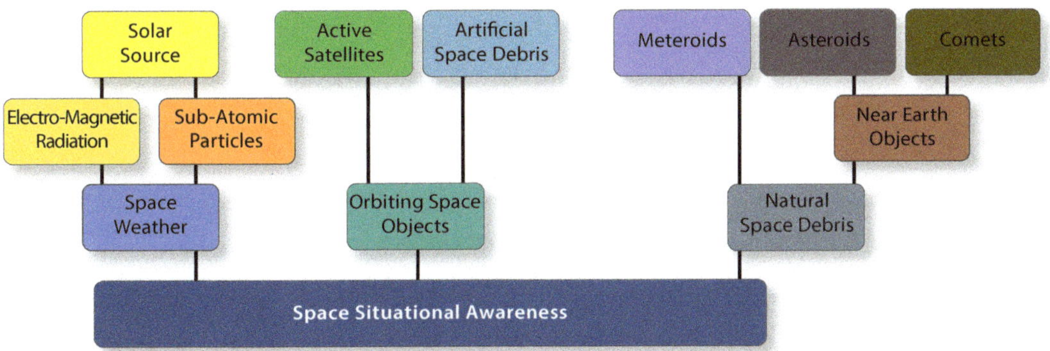

SSA can provide warnings that satellite operators should temporarily move their spacecraft to avoid a possible collision with orbital debris. The technologies can be used to alert clients to the potential for degraded data collection caused by natural events over a predicable period. These applications imply benign and cooperative activities, as some of them clearly are. But evidence indicates that SSA has several other functions.[12]

If one or more SSA sensors can detect and track the position of space debris in relation to spacecraft, then the same technology can obviously geolocate larger objects in outer space and assist in guiding space weapons toward targeted spacecraft. SSA could additionally observe and measure the capacities of spacecraft to manoeuvre, contributing to a catalogue of potential military targets in outer space as well as indicating the most appropriate strategies for attacks.

SSA as practised by the major spacefaring States is a therefore a component of outer space stability and efficiency as well as an intelligence gathering apparatus designed to support aggressive action in outer space. SSA becomes a component of major power entanglement strategies when installations are positioned on client-State territory. Investment in SSA capabilities is increasing.

And finally, as part of their defence and national security arrangements, the major States and almost all others understandably have in place arrangements for denying or inhibiting access to data and information concerning advanced-weapons designs and other sensitive capabilities. But it is not clear how effective these data-security mechanisms have been or could become. Inventions with commercial value have commonly been patented with designs registered on publically accessible records (although the details of some more sensitive technologies did not become publically available until after a period of security-related hibernation had elapsed).[13] Tertiary institutions encourage dissemination by supporting the publication of scientific papers in reputable journals. These arrangements have, perhaps unavoidably, made available a constant and expanding stream of intellectual property to an anonymous and global audience.

Industrial and related espionage appear to be commonplace and such incidents are often recorded in the media. Every reported incident of espionage may reflect adversely upon those responsible for preventing such events, so it is likely the number of occurrences is greater than has been revealed. Interactive and informal electronic networks installed on cell-phones and similar devices have also made the containment or suppression of information more difficult than it was in the past. These processes enable the distribution of classified information by disenchanted government employees, some apparently acting independently, and by others.

Export sales of armaments, while supporting entanglement objectives, may have provided opportunities for the imitation of new space-related technologies through reverse engineering. Perversely, some provider-State constraints on arms sales and technology transfers appear to have encouraged competitive and duplicative research. [14]

The implications of these developments for space warfare are threefold. First, not one of the major States can be certain its competitors have failed to duplicate a revered discovery regarding a classified weapon or related capability, or obtained key information about it through espionage or via some other avenue. In consequence, notions of superior space power based on as-yet-unrevealed capabilities, if any such capabilities exist, may be wholly or partially illusory. Second, persistent competition for technological superiority between major States increases further the probability that they will not, during the foreseeable future, act cooperatively to enhance global space security. Third, the most powerful States, spurred by their histories, national personas, politics and addiction to competitive behaviour, have no apparent alternative but to persist with guarding information in the hope of maintaining a minimally porous membrane between new technologies and the apparently continuous processes of proliferation.

Nevertheless, given the entrenched strategic and geopolitical realities referred to previously it is difficult to identify alternative approaches the major spacefaring States could at present pursue.

The next component of this exploration, *Acquire Freedom*, considers options available to the less prominent States wishing to reduce the prospect of warfare in outer space and to limit adverse consequences should it occur.

Acquire Freedom

The potential for human activity to disrupt the provision of space-sourced data and services appears likely to persist into the indefinite future. The struggle involving the US and USSR, now Russia, has subsided but not evaporated. In its place there is rising a contest for influence between the US and the PRC. The Russian-PRC relationship is peaceful but has in the recent past been strained. Other potentially potent strategic influences are emerging such as the rise of India, Japan, RoK, Israel, DPRK and just possibly Europe.

Against this background, opportunities exist or could be created enabling middle-range and smaller State powers to enhance outer space security, independent of major power ambitions. The justification for any such initiatives is acceptance of the reality that every State would be affected by a major space conflict, regardless of whether the State had directly or indirectly supported any associated military activity or had merely observed unfolding events.

However, geopolitically significant initiatives adopted by less powerful nations could be viewed by the major States as introducing new and potentially uncontrollable factors into the global strategic equation. Prominent States have in the past demonstrated a willingness to intervene when they disapprove of an emerging trend. Contemporary examples are the sanctions imposed against the DPRK and Iran for their actual and alleged involvement with nuclear weapons, although many other instances have been recorded. States contemplating innovative policies affecting space security would need to recognise this prospect and to represent intended developments as likely to enhance global stability with implications for reduced tension and other benefits, including benefits for the major States.

Also, at a minimum the major States might reasonably expect the positive and global contributions they have made through space technologies to be recognised. Prominent States, the US more than any other, have contributed to global economic growth and to associated benefits by providing space services and data as free goods. These services have included almost ubiquitous GNSS timing and geolocation data, meteorological data unobtainable other than from spacecraft, SSA data, environmental monitoring data, global search and rescue services and scientific data. This apparent altruism by major spacefaring States complicates the ethical setting in which other States might consider the pursuit of independent strategies in relation to outer space security. It also points again to the desirability of avoiding activities likely to damage powerful-State security.

But appropriate gratitude need not nullify all possible initiatives. Throughout the space-age, economic benefits have not always flowed from the powerful to the less powerful States. Smaller States, particularly those with low-cost labour forces, can claim to have contributed to the stronger economies of major powers. Additionally, the number of spacefaring States has been increasing, which has both enriched and further expanded the range of space-related services available globally. By 2009, 57 States and other entities had satellites in orbit.[15] The more recent State entrants into outer space, and States with evolving space dependencies, may consider themselves entitled to a more effective voice in matters concerning security and the deterrence of warfare in outer space. Additionally, the altruistic sharing of space services and data by major spacefaring powers arguably led to some domestic benefits for themselves.[16]

The situation now confronting smaller and non-spacefaring States is in some respects analogous to the challenges experienced by landlocked nations in relation to law concerning the high seas. Rights of access by landlocked States to the high seas were recognised at a conference held in Barcelona during 1921 where there was agreement to a *Convention and Statute on Freedom of Transit* (Barcelona Statute) and to the *Declaration Recognising the Right to a Flag of States having no Sea Coast* (Related Flag Agreement). The rights of landlocked States are now contained in Part X of the *United Nations Convention on the Law of the Sea* (Law of the Sea Convention). Sanger, 1987, has recorded the persistence and influence exerted by Latin American and African nations when aspects of the Law of the Sea Convention were being negotiated.[17]

There is scope for the smaller States to again act effectively in both proposing and implementing ways to reduce the threat of warfare in outer space.

Many of the middle-level and less powerful States could work cooperatively on programs based on their common interests and under conditions effectively not available to the established major powers. Most of these other nation States are similarly competitive in extraterritorial arrangements and ambitions. But most are not committed by strategic history or political pressure to maintaining inventories of strategic weapons, to seeking global authority, to viewing themselves as world leaders or to projecting their own political and economic policies globally. Their strategic interests are typically, although not exclusively, regional.

If these States want a stronger voice than they now have in international affairs they will require a mutually recognised forum for debate and a platform from which treaties and other determinations binding upon them could be promulgated. Any such forum would almost inevitably and appropriately adopt agenda extending beyond security in outer space.

The potential for the smaller States to act in concert does not assume all or many cooperating nations working compatibly as a unified bloc on every initiative that might be proposed. Differences in culture, economic objectives, strategic geography, financial capacity, and international ambition may create divisions. There are nevertheless grounds for broad agreements on some or many issues, particularly those affecting outer space. In consequence, participatory opportunities would be available to almost every State. And the potential for obtaining agreement to courses of action proposed would likely be significantly greater than when the five diverse but permanent members of the UN Security Council attempt to find common ground on almost any topic.

Precedents exist for successful arrangements operating in parallel with UN-sponsored initiatives. They led to, for example, the Chicago Convention, administered by the International Civil Aviation Organisation (ICAO) which only subsequently became an agency of the UN, the Antarctic Treaty Consultative Meetings, and the formation of NATO. The Open-ended Informal Consultative Process on Oceans and the Law of the Sea is an example of widespread support for the periodic review of an existing treaty.

There also are numerous examples of organisations being able to endure or persist despite periods of both peace and war affecting participants. The First International Meteorological Conference was held in 1853. The committee which became the International Committee of the Red Cross was formed in 1863. The Universal Postal Union emerged in 1864 and the predecessor to the current International Telecommunications Union (ITU) in 1865.

Harding, 2013, identified seven regional organisations with a space focus, indicating a nascent recognition by some States that new and alternative mechanisms have potential to address and progress issues related to outer space.[18]

Multilateral negotiations, parallel to the UN, in matters specifically related to outer space have already occurred. The European Space Agency and European Union have attempted to develop an outer space code of conduct aimed in part at reducing space debris. On 17 January 2012 the US announced it would join this consultative process involving the European Union and other nations with a view to drafting an international code of conduct for outer space, although some or all of those initiatives might have been diluted or negated by the work of a *'group of governmental experts'* which concluded in July 2013. A proposal for selected States to enter an agreement for protecting historically significant sites on the Moon has been informally proposed.[19]

If a large number of States became involved in the forum envisioned, decisions taken could gain international recognition beyond immediate signatories. Provisions allowing for subsequent membership increases and for accessions to proposals already agreed would potentially result in further expansion of influence and of commitment.

In all such arrangements a one-State-to-one-vote constitutional principle would appear central to success and would act as a safeguard against manipulation or paralysis by a minority of States members claiming to require a consensus process, as apparently happens with COPUOS.

Effective leadership would seem necessary to prioritise and coordinate activity in any such forum and to offer logistic services, particularly during an establishment phase. The prospect of a leadership vacuum would open an opportunity for an influential State to take a prominent role. This could enhance its global leadership status, portray it as committed to progress in outer space and demonstrate its concern for enhancing space-related security. Such a State could also project itself as concerned with advancing the rights of less-prominent or more vulnerable nations. In short, there exists potential for a significant political coup, a diplomatic masterstroke with global ramifications.

Candidate States for the initial leadership function would include India, Brazil or possibly Canada, which has a strong record of seeking international stability, including in outer space.[20] US participation with other than observer status seems improbable, taking account of its consistent position in matters related to new or additional constraints on outer space security-related issues. The PRC or Russia would be somewhat more likely to engage with such a forum.

Even if all three of the major spacefaring States boycotted the new forum, that need not be fatal. As recorded at Chapter 7, Table 7 – 1, *Selected International Agreements*, Russia the US and the PRC are not signatories to the Anti-Personnel Landmines Convention, or to the Convention on Cluster Munitions. As at 27 December 2012, 133 States had signed and a total of 161 States had agreed to be bound by the Anti-Personnel Landmines Convention. By 4 March 2013 a total of 111 States had joined the Convention on Cluster Munitions. Additionally, as at January 2013 the Law of the Sea Convention had been ratified by 165 States (total including the European Union) but not by the US. Consequently, a refusal by one or more of the major States to participate either fully or in principle, would not necessarily prevent significant progress being made through international agreements. Exceptionalism has its advantages but also camouflages some unique risks.

If the decision was taken by a consortium of smaller States to develop such a forum, alternative to and in parallel with the UN, their potential objectives would need to recognise the comprehensive and beneficial services extended globally by the United Nations through its network of committees, agencies and related bodies including ICAO and the ITU. This implies a bounded coalition arrangement, but one sufficiently broad to achieve required outcomes.

There already exists one precedent, and possibly more, where a UN Secretary-General has reached a cooperative agreement with a parallel security-related organisation, namely the *UN-NATO Cooperative Declaration* signed on 23 September 2008 by UN Secretary-General, Ban Ki-Moon.

A potent incentive to participation in such a forum is that the majority of States have been deprived of meaningful influence over space security, with implications for their own national security. A *Prevention of Arms Race in Outer Space* (PAROS) treaty proposal was submitted by the USSR to the UN Conference on Disarmament in 1981. Two ostensibly related but diverse resolutions were then adopted by the UN General Assembly on 9 December 1981. The first resolution (A/RES/36/97) proposed a negotiated and verifiable agreement to prohibit anti-satellite systems. The second resolution (A/RES/36/99) proposed negotiation of a treaty to prohibit the stationing of weapons of any kind in outer space.

A combination of the USSR proposal and the two related Resolutions led to almost two decades of unproductive negotiation. Canada in 1998 attempted to resuscitate the debate by suggesting an *ad hoc* committee be formed to negotiate a convention for the non-weaponisation of outer space. Canada renewed its proposal in 1999 and then submitted additional papers in 2006, 2007 and 2009 all without apparently measurable progress being achieved.

In February 2000 the PRC reiterated the need for an ad hoc committee to address PAROS and in 2002 Russia and the PRC jointly submitted to the Conference on Disarmament a paper attempting to advance this issue. In 2008 Russia and the PRC submitted a revised proposal in the form of a draft treaty on the prevention of placing weapons in outer space without discernible progress being made subsequently. Between December 1958 and December 2011 the UN General Assembly adopted a total of 111 space-related resolutions, but apart from resolutions adopted to endorse treaties already agreed between the major States, evidence of those resolutions having been influential is slight.[21]

The following Table 8 – 1. *UN Voting on Outer Space Security, Every Third Year,* summarises voting on the PAROS and related proposals. Data for each of the remaining two-year sets is consistent with that shown in the table.

Table 8 – 1. UN Voting on Outer Space Security, Every Third Year [22] [23]

Year	Yes Vote	No Vote	Abstention	Non-voting	Total Voting Membership
2009	176	0	2	14	192
2006	178	1	1	12	192
2003	174	0	4	13	191
2000	163	0	3	23	189
1997	128	0	39	18	185
1994	170	0	1	14	185
1991	155	0	1	10	166
1988	154	1	0	4	159
1985	151	0	2	6	159
1982	138	1	7	11	157

Referring back to Table 6 – 1, *Ratifications and Accessions to Outer Space Treaties,* it was clear that despite constant or heightened concern over outer space security and stability, as shown by the table immediately above, only 102 States have ratified or acceded to the 1967 Outer Space Treaty. The number of ratifying or acceding States had declined with every subsequent treaty, only 15 States having thus endorsed the 1984 Moon Agreement. Given the history of voting in the United Nations on issues affecting outer space security, this was not a surprising outcome. Smaller States can see little or no purpose in subscribing to a legal architecture designed and maintained by States over which they have minimal influence.

One possible illusion of progress emerged in 2010, under the auspices of the UN, when the group of '*governmental experts*', previously mentioned, was formed to conduct a study on '*outer space transparency and confidence-building measures*'.[24] The group submitted its report to the General Assembly on 29 July 2013. This report was presented in broad terms allowing significant scope for interpretation or reinterpretation as appropriate to various States. Only 15 States were represented on the group, although those States included Russia, the PRC and the US. All recommendations made by the group were '*non-legally binding voluntary*'.

It is apparent from these examples, with regard to the development of the international law affecting the security of outer space, the UN with its associated committees and agencies has failed to meet the expectations of the majority of its States members. There are few if any indications of the situation changing within the UN.

Most States members of the UN now face a stark choice. They can accept they are, and will remain, almost impotent in relation to advancing the security of outer space. This may have adverse consequences for their own national security and economic development. Alternatively, they can begin to pursue initiatives through arrangements alternative to the United Nations, conceivably through an organ as conceptualised above

A coalition of States could agree to develop for example:

a. Enhanced versions of the existing space treaties.
b. Clarification of limits to the vertical sovereignty of States.
c. Greater specificity of sovereign activities permitted in outer space.
d. Increased constraints on weapons and on hostile activities in outer space.
e. A more detailed and binding space code of conduct if other attempts to produce one should either fail or become diluted by negotiation until they are almost meaningless.

Decisions taken by only a few of the less prominent States could lead to a failed outcome in that their determinations would be ignored by most other nations. Conversely, even single-State pronouncements can also have regional or wider ramifications.[25] One example of an effective single-State pronouncement which eventually had global implications was the US's *Policy of the United States with Respect to the Natural Resources of the Subsoil and Seabed of the Continental Shelf* (the Truman Declaration).

The support of many States would nevertheless endorse and legitimatise claims to have created new, visible and influential international values and standards. Given the voting records of most States on PAROS for example, it is likely a forum such as that described above would incorporate a large number of disaffected nation States. Declarations within the forum could be expressed as multilateral treaties, binding upon participating States, with further potential to metamorphose into aspects of customary international law.

Negative comparisons might be made between determinations as proposed and the allegedly failed *Declaration of the First Meeting of Equatorial Countries* (the Bogotá Declaration, 1976). In that instance, eight equatorial States (Brazil, Colombia, Congo, Ecuador, Indonesia, Kenya, Uganda and Zaire) declared the GEO orbit to be part of their sovereign territory. The Bogotá Declaration was consistently ignored by nations other perhaps than its signatories.

However, the Bogotá Declaration established a political precedent. A group of States other than the most prominent nations had formulated a proposal regarding outer space. They had shared common ground only in that their sovereign terrestrial territories were located on the equator. Concerns regarding outer space had united and motivated them to agree on a shared policy. The eight States lacked capacity to ensure their claim was comprehended and acted upon globally, but they pointed to a potential for smaller States, when united, to pursue interests of mutual concern, just as the smaller and landlocked States had done when the Law of the Sea Convention was being formulated.

There are also at least two fundamental distinctions between the Bogotá Declaration and potential space-related declarations able to be made by a consortium of States acting in parallel with the UN.

a. The Bogotá Declaration was made in opposition to established international law. By 1976 satellites had been in the geosynchronous orbit for approximately 13 years without visible protest or condemnation. The Outer Space Treaty declaring outer space to be a region free from any claims of sovereignty had been in force for nine years. In contrast with the Bogotá Declaration, actions by a coalition of States as envisioned in this work would be clarifying, consolidating and developing established international law in ways broadly compatible with existing treaty provisions.

b. States signatories to the Bogotá Declaration had minimal technical capacity to project their alleged authority 36 000 km into space other perhaps than by jamming the transmissions from communications satellites. By contrast, some States already have a capacity to deploy defence technologies to an altitude of 80 km, the lowest boundary to outer space as proposed by Appendix 1. Defensive activities in that context would protect contiguous sovereign territory.

Entanglement arrangements between States were considered earlier. The paragraphs following address only those entanglement or alliance issues related to the establishment of military bases and related weapons systems by major powers in weaker foreign States.

The major States already have extensive strategic strike capacity enabled by weapons available for release from their own territories or from mobile platforms such as aircraft, submarines or UAVs. But by foreign basing they are able to distribute tactical and other weapons yet more widely while also exporting the territorial risks and consequences if facilities are attacked Every such foreign installation brings with it the collateral benefit that it is a symbolic expression of extended major-State power and influence.

Benefits to client States hosting foreign installations can include associated security agreements which might become valuable *if* they are both needed and then honoured, neither of which cannot be counted on. Disadvantages to client States are also identifiable, particularly when basing of missiles is involved. A commonly claimed justification for such activity is protection of the client State from attacks which, impliedly, could not otherwise be countered or defeated.[26] Questions not always addressed relate to whether a hypothesised attack would occur if the foreign base or weapons system had not been installed, or if an attack was reasonably likely *per se*. Client States assume additional risks of pre-emptive attacks when they accept foreign surveillance radars, military ground stations or multi-purpose SSA facilities.

These considerations are relevant to potential warfare in outer space. US-encouraged and enabled missile and other defence arrangements appear to be resulting in the partial encirclement of Russia and / or the PRC. Under the US '*phased adaptive approach*' to the installation of missiles in Europe, installations will include the SM-3 missile, a variant of which has been used to intercept and destroy a satellite. That missile is scheduled to be either replaced with, or possibly supplemented by, a yet more capable missile now under development by the US and Japan jointly, the new SM-3 Block- IIA missile. Russia has objected, apparently ineffectually, to the missile basing proposals. The US has been reported as accelerating installation of missiles in the Persian Gulf region.[27]

There may be opportunities for extant or potential client States to reject foreign basing of weapons and personnel or to negotiate agreements more directly supportive of their own strategic interests. To provide for minimal erosion of their sovereignty and to maximise independence of action, client States could when politically possible insist upon domestic control or influence over when, how and against whom the weapons proposed to be on their soil could be used. Where space and other surveillance facilities, ostensibly intended for benign applications, are proposed then the client States could seek maximum involvement, multinational staffing and verifiable access to data collected.

Potential client States also have the option of avoiding or renegotiating bilateral treaty agreements in which outcomes could be influenced by the significant negotiating resources available to a major State. Multilateral negotiations provide client States, acting in concert, with opportunities to utilise additional diplomatic resources while forming closer relations with compatible entities. This approach would probably not reflect powerful State preferences. The PRC has in the past favoured bilateral arrangements, having between 1973 and 1982 entered into 1 395 of them, although I did not locate more contemporary figures applying to the PRC's treaties.[28] Bilateral arrangements are also preferred by the US which, as at January 2011 had bilateral treaties with 232 nation States and their colonies. [29] I did not find comparable statistics applying to Russia.

Most of the smaller States have no viable alternative to relying on some data and services provided by space assets they neither own nor control. They realistically perceive the technological, materiel and financial investment needed acquire a spectrum of indigenous satellites and associated sensors as being beyond their capacity. They therefore enter arrangements to acquire data and services either from spacefaring States or related commercial entities.

State-supplied data and services are not necessarily reliable. States, particularly the most prominent of them, have technical, diplomatic and financial capacity to selectively limit or deny access to space-sourced data, one example of this being the US severing intelligence ties with New Zealand as already mentioned. During periods of military conflict a provider-State may refuse to share information on territories hosting military initiatives and troops, particularly if a client State is not involved directly as a participant in the military operation. Entering arrangements with a major State power as an almost sole source provider of space data and services can additionally generate obligations of compliance with, or overt support for, provider-State strategies while curtailing operational sovereignty. This would predictably limit options for a client State to press for increased security in outer space except in accordance with provider State preferences.

Dealing with foreign commercial organisations operating satellites, assuming they could provide data and services of an acceptable standard, might not be more reliable than reaching agreements directly with foreign State governments. The commercial environment creates its own risks. A commercial operator might sell a spacecraft to another organisation under circumstances making it difficult or impossible to enforce an original service-provider agreement. A company could fail in business with assets reverting to financial or other institutions interested only in recovering a capital investment regardless of prior arrangements. In the case of an accident, damage in warfare, or technical failure affecting a satellite already on orbit, a commercial organisation might not be able to raise capital for building a replacement spacecraft. Even when capital was available a significant gap in the provision of services or data could be experienced. In times of tension wealthy States can and do exert pressure on commercial operators to enter exclusive client agreements which can result in denial of product to the other customers.[30]

It is not possible to generalise over whether State or commercial assets in outer space are either more or less prone to attack or interference. In some circumstances spacecraft belonging to commercial entities dealing with a spectrum of international clients may be less prone to attacks in outer space than are State-owned assets, essentially because the consequences of aggressive action would attract adverse responses from a range of participating States. Conversely, attacks against commercial satellites might have greater prospects of success than attacks against military spacecraft because military spacecraft are likely to be better protected by design and construction. Commercial spacecraft, particularly those supporting military operations, might thus become preferred targets notwithstanding associated disincentives.

Taking account of these issues, the interests of client States would appear best served by avoiding sole-source commitments and spreading space-related acquisitions of data and services across two or more provider States as well as across power blocs and corporations in competition with each other for markets and for influence. Specific sensors and components of larger spacecraft could also be purchased outright with ownership of a suite of sensors distributed across several corporations or across various States.[31] [32]

It is demonstrably practicable to form data and service supply relationships with diverse vendors. India, Brazil and Indonesia are among the States which have already done so. Adoption of a non-aligned supply posture may lead to exclusion from the most recent technologies developed by one or other of the major powers. However, as considered in Chapter 1, access to the most recent technologies does not guarantee enhanced national security.

For reasons already considered it is difficult or impossible for powerful States to exert total control over the intellectual output of their universities, research institutions and other sources of innovation. For many other States this incapacity may prove to be advantageous.

Global dissemination of data and information relating to space weapons and related technologies should strengthen potential outer-space combatants to a comparable extent thus limiting the capacity of any one of them to become technologically dominant or to project the image of having achieved dominance.[33] Under those circumstances potential success in space warfare is likely to become even more elusive than it is already.

Less powerful States therefore might, to the extent strategically and politically possible, adopt a policy of minimally constraining publication of their own sensitive technological information concerning space-related weapons and similar technologies. At a minimum the smaller and non-aligned States have available the option of refusing to sponsor research likely to contribute to the development of space warfare and weapons. Some, however, will also have capacity to sponsor research contributing to the survivability of spacecraft operating in hostile environments.

There already exist opportunities for research involving, for example: designing enhanced protective sheaths for spacecraft, improving spacecraft profiles to resist impacts, improving radiation hardening techniques, designing new and more economical redundancy circuits, increasing immunity to electronic attack including through better antennae designs, inventing improved signal encryption and jam-resistance techniques for radiofrequency signals, exploring innovative techniques for deorbiting or otherwise reducing existing space debris, improving thermal protection for spacecraft components and enhancing the capacity of imaging apertures and of solar-energy panels to resist laser or related attacks.

As with most aspects of space-related technology, the unconstrained release of any such research could have unanticipated consequences and would therefore need to be approached cautiously. If, for example, satellites could be made less vulnerable through enhanced electronic defences, then so in some circumstances could orbiting weapons, including co-orbital satellites, be made more survivable. Additionally, the higher the level of protection from electronic and related attacks against satellites, the greater could be the temptation for an aggressor to use kinetic or other weapons likely to cause debris clouds. Consequently, decisions to adopt this course would need to be taken after careful review of the broader implications on a case by case basis.

The concept of territory over which no State can exert total control, presently applicable to outer space, the high seas, and airspace above the high seas, has been expanding subtly to include the virtual territory of economic trade between sovereign States. The large economies of powerful States in particular are getting increasingly entwined as they exert and are vulnerable to international fiscal, monetary and other economic influences generated by each other. The 2011 percentages of selected State GDP attributed to merchandise trade, for example, were: US 24.8; Japan 28.6; Russia 45.5; India 40.5 and PRC 49.8.[34]

Economic interdependence is likely to exert a leavening influence on international relations. The influence will obviously not be strong enough to eliminate the ingrained competitiveness of nation States, but should be sufficiently potent to modify it. Economic globalisation could therefore contribute to international stability with potential to reduce tensions in outer space as in all other strategic environments. Smaller States might therefore consider this factor when determining their policies in relation to globalisation.

Possibly modifying this beneficial potential is that State-based space industries, although subject to globalisation pressures, confront influences incompatible with responsive free-market principles.[35] The US and some other States have, for example, imposed standards and export restrictions on space-related technologies, causing satellite manufacturers to object that new business is being captured by companies in the PRC, India and other States which market less expensive products. Conversely, during the period June 2005 to December 2012 the global space-stock index closely tracked both the Standard & Poor's 500 (S&P 500) and the National Association of Securities Dealers Automated Quotations (NASDAQ) indexes. This result suggested governmental intervention in Western space industries might not have impacted corporate profitability to the extent claimed by some business executives.[36] Additionally, captive markets in the form of home-State government contracts can have compensatory effects.

In the event of a conflict in outer space, the least protected satellites and related space vehicles may be most vulnerable to intentional harm and *vice versa*. Sellers in the space marketplace can therefore alert clients to the potential for conflict in outer space by advertising advantages associated with higher standards of security and reliability. Eventually, financial pressure combined with any tendency for one or more of the powerful States to lose market share and, possibly more importantly, to forego some international influence, should lead to approximate market equilibrium.

In summary the smaller States will remain disenfranchised if they continue to pursue futile attempts to promote enhanced security and stability in outer space through the currently established diplomatic and consultative mechanisms. Other avenues and options can be created by them for the ultimate benefit of all nation States.

Share Freedom

All States benefit from stability in and continuity of access to outer space-sourced technologies, data and services. This global reliance offers opportunities to enhance space security through initiatives upon which the overwhelming majority of States, large or small, weak or powerful, could agree. These steps could be taken without affecting most extant bilateral and other entanglements.

The prospect of false alarms, indicating that attacks in or from outer space are imminent, is not a theoretical or esoteric issue. Sennott, c. 1987, cited papers released under the US Freedom of Information Act reporting a total of 1 152 moderately serious false alarms affecting the US during the period 1977 – 1984.[37] These Cold War-era alarms were generally resolved by a process of dual phenomenology requiring confirmation of an attack from two separate sensor technologies.

But not all such incidents have been confined to either the Cold War or to situations initiated by major spacefaring powers. On 25 January 1995 a Norwegian sounding rocket was misidentified by Russian radar as a Trident missile.[38] In March 2012 Japan threatened to shoot down a DPRK vehicle which might have been either a missile under test or a space-launch vehicle, and which might or might not have entered Japanese sovereign territory. [39] During 2013 – 14 a series of incidents in the US indicated that supervision of its ICBM fleet was sub-optimal, although arrangements in other States were not necessarily better, albeit their weaknesses if any were not disclosed.[40]

The risk of a false alarm pointing to an imminent attack within outer space is mounting. The number of satellites and satellite operators is increasing.[41] Spacecraft operators have access to new technologies, such as powerful thrusters, enabling vehicles to rapidly approach other objects in orbit thus potentially causing concern regarding malevolent intentions. Some States, now armed with intermediate-range missiles, might respond to false alarms with less provocation, less data, and therefore less certainty than has in the past been required by major spacefaring States. The cost of construction, installation and maintenance of dual phenomenology architecture will preclude smaller States obtaining that capability in the foreseeable future. The adequacy of some existing national early warning technologies has at times been questioned.

It would support outer space stability if all affected States ensured procedures relating to space launch and related activities under their control could not be misinterpreted. The manner in which this could be achieved would vary in accordance with the activity proposed. It would need to allow for the possibility of a launch or other space-related activity failing to function as intended.

Regardless of specific arrangements adopted, one consistent requirement of enhanced early warning will be the maintenance by all spacefaring States of a highly reliable communications network, incorporating all other spacefaring States, so that intended activities could be advised in advance and any detected deviations or anomalies explained without delay. A capacity for immediate responses to inquiries from any State agency would be an indispensable component.

The issue of avoiding false alarms by an international constituency has already been addressed in part by the *Hague Code of Conduct Against Ballistic Missile Proliferation.* Clause 4 (a) (iii) of the Code provides for subscribing States to exchange notifications on their ballistic missile and space vehicle launches and test flights. However, the Code does not extend to all terrestrial situations with potential to cause false alarms. It would not have prevented the Norwegian emergency cited. Neither does it make provision for notifications of planned changes to spacecraft orbits or to other spacecraft manoeuvres conceivably leading to conjunctions with other spacecraft and therefore able to trigger alarms. Bilateral arrangements between States to avoid false alarms cannot provide for the broader and uniform participation of many other States with a potential to create destabilising events.

All spacefaring States, regardless of whether they already have in place specified arrangements to address this issue, could adopt appropriately comprehensive and standardised protocols. These activities might on request be guided by advice or technical support from the US, Russia or the PRC, functioning either individually or cooperatively. A new treaty formalising agreed arrangements would support this aspect of security in outer space. The report of the group of government experts into outer space transparency and confidence-building measures, already mentioned, might also be interpreted as supporting the notion of an effective arrangement for avoiding false alarms. Potentially relevant concepts were included in their submission. [42]

It will be challenging to convince States to consider voluntarily reducing reliance on space services and data, as insurance against an unmeasurable potential for warfare in outer space. Outer space dependencies in some States have reached extreme levels, and are embedded throughout social, commercial and national security infrastructures.[43] Those data and services are perceived as real assets whereas the potential for warfare in outer space could conveniently be dismissed as remote or merely hypothetical, regardless of evidence to the contrary.

Additionally, the global economic benefits of utilising data and services obtained from outer space might never be known, in part because the utility and value of space support is constantly changing, in part because financial investment in space-related initiatives can vary significantly between years, and in part because the quantification process is itself exceptionally complicated. Published estimates, commonly not supported by reliable sources or explanations of methodologies, tend to agree that global space-related expenditure in 2012 was about USD 304.31 billion and for 2013 approximately USD 314.0 billion, although expenditure by *governments* on space fell during that latter year.[44] The returns on investment in space-related activities vary but are commonly positive as implied by space industry stock movement, particularly in relation to S&P 500 and the NASDAQ, as recorded previously.

There is, however, potential for the exceptionally costly and valuable space-sourced data and services to be terminated, interrupted or degraded in quality by aggressive State activity, provider-intervention, natural events, technological failures, corporate collapses or interference by non-State activists. On some occasions the cause of such an event cannot be identified, at least in the short term, an example being on 2 April 2014 all 24 satellites comprising the GLONASS global satellite navigation system failed simultaneously in an outage lasting 13 hours.

The extent to which space-sourced data and services are duplicated, supplemented or substituted by other means will be influenced by each State's awareness of the costs and consequences of either partial or total space denial, and by each State's financial, technological and political capacity to insure itself, noting some initiatives described below could be financially cost-effective.

Some available methods of reducing space reliance are prosaic but nevertheless potentially valuable. They include, for example, ensuring civil and military aircrew remain competent to navigate aircraft without support from GNSS, and rehearsing both national security and emergency service personnel in functioning without support from space. Tactics of this nature may need regulatory action potentially supplemented by a modest investment in training.

Military acquisitions can be biased toward equipment with a capacity to function without space support in some instances. Weapons systems can include precise munitions with dual guidance systems such as radar, laser, heat-seeking warheads and other target acquisition technologies in addition to GNSS.[45] Terrestrially controlled UAVs can be equipped to relay beyond-line-of-sight communications as a temporary measure in the event of short-term denial of access to some space services.

For national security, military and commercial applications, a terrestrial substitute technology for GPS has already been developed by an Australian company, the Locata Corporation. The Locata system provides positioning data as accurate as is available from GPS, the equipment can be positioned to avoid terrain masking and Locata is equipped with variable signal transmission strength (RMS) able to defeat most of the radiofrequency jammers that can block conventional GPS signals.

The Locata option might appear unpalatable because it involves an initial financial outlay when at present GPS is a free service. However, installations of a substitute for GPS, or GNSS more generally, could be limited geographically to areas of vital strategic significance. Locata-related costs could be met from user-pays arrangements in some cases. The national security and economic benefits of having this technological insurance available are incalculable but potentially significant.

Other alternatives to reliance on some aspects of GNSS are also emerging. They include technologies based on refinements to inertial navigation and on new approaches to celestial navigation.

Not all terrestrial alternatives to space data and services require investments justified solely by the possibility of space denial. Radiofrequency spectrum is a finite resource and one under constant pressure. With increasing demand, regardless of which technologies are invented to compress and sequence transmissions, the point will be reached where communications reliant on the radiofrequency spectrum must become limited in some ways. States or consortia of States could forestall these developments or prevent the problem from arising while simultaneously reducing reliance on space-based services by increased investment in fixed-line communications infrastructures.

Enhanced terrestrial fixed-line and related communications assets would offer States more direct control over aspects of their critical national infrastructures than is achievable from reliance on space-sourced data and services. Following damage or disruption as a consequence of intentional action or natural events, terrestrial damage could normally be accessed and repaired. On-orbit repairs to space assets remain beyond the capacity of most States and could be denied to every State if access to outer space became unacceptably hazardous because of NDRE or a substantial increase in the density of orbiting debris.

There may also emerge opportunities to utilise space-sourced data that does not emanate from artificial satellites. One nascent example is the space navigation technology being explored by the European Space Agency and based on X-ray emissions from pulsars.[46] The potential for terrestrial applications enabled by such emerging opportunities cannot yet be confidently assessed. In most cases the utilisation of these evolving technologies will be critically dependent on a capacity for uncorrupted electromagnetic energy to traverse near-Earth space. If near-Earth space became polluted following ionising and other processes induced by anthropogenic activity, terrestrial reception could become unreliable.

In summary, numerous terrestrial alternatives are now available to duplicate or substitute for some space data and services. The range of options appears likely to increase.

Given deliberations in preceding chapters, the potential for broad international acceptance of a proposal to declare any part of outer space to be a zone of non-aggression seems negligible and the notion absurd. In the past, such initiatives have not progressed past generalised assertions such as, '*....recognising the common interest of all mankind in the progress of the exploration and use of outer space for peaceful purposes...*' (prologue to the Outer Space Treaty, 1967). The PAROS proposal has remained in limbo for more than 30 years. There is, however, one definable zone of outer space which could nevertheless be corralled by international agreement from attacks using any form of weapon or interference whatsoever.

Data and services available from GNSS spacecraft have global application and there is global dependence upon them. Primary satellite constellations comprising elements of the main GNSS architectures orbit in the same MEO band. It is, broadly, located at ~19 000 – 21 000 km mean altitude but could be further protected by higher and lower buffer zones with additional adjustments to allow for actual orbital profiles. A comprehensive and international declaration banning the presence of all forms of weapons, attacks by all weapons, and all forms of electronic interference could be agreed internationally for this specific region of outer space.

There is no obvious reason for such a proposal to receive other than almost global support, including from each of the most prominent of spacefaring States, they having the most to gain from such an initiative as well as the most to lose if a conflict involved the destruction of GNSS spacecraft.

International agreement to declaring the MEO band a zone of non-aggression would not inhibit electronic jamming or spoofing used in *terrestrial* warfare to deny GNSS reception to opponents. That capacity could persist under the arrangement proposed.

An agreement as suggested would not provide absolute protection for GNSS satellites. The GNSS spacecraft could be affected by NDRE following a nuclear detonation almost anywhere in near-Earth outer space. A State might elect to ignore or to violate the applicable international law and attack assets in the MEO orbital band. Nevertheless, as with all widely acclaimed international agreements, the arrangement would function as a disincentive. It would provide a legal basis for opposition to an offender's actions and conceivably a justification for multinational retaliation. It would also offer a foundation for beginning to use international law to further enhance the security of all assets in the outer space environment. It is practicable and achievable.

(After this proposal had been written an announcement was made that Russia and the US were seeking special arrangements for their GNSS spectrum through the UN International Committee on GNSS (ICG). However, ICG decisions are made by consensus.[47])

All of the potential global initiatives described would involve extensive multinational participation and cooperation. Cooperative implementation would emphasise the need for global commitment to reduce the prospect of violence in outer space. It would also create an environment conducive to adopting new and additional cooperative measures extending beyond outer space affairs.

This is more than vague ambition. Research has been undertaken demonstrating the potential to both build and consolidate trust under a range of political and commercial circumstances.[48] As already recorded, organisations including Red Cross, the Universal Postal Union and the predecessor to the current ITU continue to serve humankind despite extensive social and technological changes, numerous wars and fundamental developments in international law. The precedents for long-term and positive change do exist.

Conclusions

The concept of *power*, including space power, is imprecise and complicated. There are no precedents to indicate the likely progress of a significant conflict in outer space; the potential effectiveness of some space weapons is unknown and unknowable; thus participants in space warfare face more uncertainty than would combatants in most terrestrial conflicts. Notwithstanding these complexities the major spacefaring States and several others have been developing the capacity to engage in outer-space warfare, commonly in ways reflecting their terrestrial experiences.

A fundamental strategy of the most powerful States has been to maintain their pre-eminence and preserve their allegedly dominant status, in which context outer space has become but one potentially contested environment among others. These strategies are oriented toward national instead of global security, although there are signs they might not ultimately serve either objective.

The major spacefaring States claim, or imply, they have a credible capacity to respond to attacks in outer space; but the notion of a capacity to retaliate in outer space, credibly or otherwise, is more complicated than State policy pronouncements imply. Powerful States thus appear to be functioning within a zone of uncertainty; their apparently concrete policies and initiatives rest upon insubstantial foundations. However, within limitations imposed by their economic, political and strategic perceptions, supplemented by an incessant drive to expand or consolidate influence in a competitive environment, the major powers may have no alternative but to persist with extant strategies and policies. Only minor modifications might be achievable. This indicates a requirement for unique initiatives by a new entity, one with different values, if the prospect of space warfare is to diminish and be contained through enhancements to global security.

There exist coherent foundations from which many of the middle-level and less powerful States could work cooperatively toward enhancing security and stability in the outer space region under conditions effectively not available to the established major powers. The majority of these smaller States have been deprived of meaningful influence within the UN forums. Precedents exist for the creation of an alternative forum for review of space-security related issues and for the implementation of multilateral treaties attracting widespread endorsement in parallel with the UN mechanisms.

In addition, the middle-level and less powerful States have options and precedents for becoming less aligned with any of the established power blocs, thus permitting increased freedom of policy and action. Some of these States could support enhanced space security through directing their national research toward, for example, the protection of spacecraft irrespective of State ownership. They could also, as one component of a strategic realignment, adopt distributed supplier arrangements for national security equipment, data and services, as some other States have already done. By supporting global economic interdependence the less-aligned States would be endorsing a leavening influence on international competitiveness.

In addition, there is both capacity and justification for all States, large or small, weak or powerful to join in supporting creation of a zone of non-aggression to enhance protection of GNSS spacecraft. This would act as a precedent for reorienting outer space issues more toward globalised considerations. Enhanced cooperative action is also practicable for avoiding destabilising false alarms associated with space launches and manoeuvres in outer space. Technological solutions aimed at reducing dependence on outer space data and services could be adopted. These global strategies offer opportunities to enhance space security through world-wide confidence building and cooperation.

Auspiciously, the fact there exist several States technologically capable of conducting space warfare at various levels and using arrange of technologies, indicates that hegemonic ambition by any single nation State toward the outer space region is absurd.

Chapter 9. Reflection

The potential for warfare to involve outer space has been recognised for decades and numerous texts on the subject have been published. However, most researchers considered events from the 20th century as a basis for their studies. Typically they examined the capacity of one nation State, commonly their own, to successfully pursue such a conflict.

Chapter 1 of this investigation took a fundamentally different course. Historical and strategic landscapes were reviewed from a global perspective devoid of nationalistic inclinations. Organised violence across 12 000 years was examined to disclose deeply entrenched behaviours and attitudes affecting contemporary activity in outer space. This pointed to actions and attitudes likely to persist unless the conduct of nation States regarding outer space can be modified.

The next following chapters considered claims made regarding weapons designed to be used in outer space. Weapons proven to have been effective when deployed against spacecraft were identified together with other weapons apparently unlikely or less likely to meet this objective. Of particular concern was that the weapons most likely to inflict catastrophic damage on orbiting spacecraft would also create clouds of orbiting debris. Even the overtly non-fragmenting, destructive technologies could trigger salvage fuses potentially carried by target spacecraft resulting in yet more orbiting detritus.

Chapter 6 reviewed international law, with emphasis on its implications for conflict in outer space. This chapter concluded the law, while providing a useful regulatory framework with some persuasive authority, was but a fragile deterrent to the outbreak of a conflict in outer space.

Providing a background for the geopolitical environment in which any attempt to discourage or curtail warfare in outer space might be conducted, Chapter 7 considered the framework in which nation States commonly conduct their extraterritorial affairs. It then recorded a vignette of influences affecting policy formulation within the more prominent spacefaring States.

Emphasising the need for decisive action are indications that, despite demonstrable capacity within several States to *conduct* warfare in outer space, it is less apparent any State could *win* such a conflict, much less control or manage one, although the potential for enduring global disruption following it remains incontrovertible.

Building upon the preceding argument, Chapter 8 proposed a range of strategies designed to enhance the security of all nation States by reducing the likelihood of warfare in space or, if it occurs, by limiting the harm it would bring to every State.

Several proposals involved forms of jurisprudential intervention based on further development of international law, despite its identified fragility. In this context, actions by some prominent States, emphasising their support for the rule of law, pointed to their perceptions of value in projecting a law-abiding image. If the image remains a desirable asset then the potential for it to be curtailed or undermined will be of concern. This vulnerability provides one foundation for the jurisprudentially oriented options to enhance security in outer space.

However, given the record of States either ignoring international law or reinterpreting it to meet indigenous agendas, it was apparent that revisions to international law could not in isolation enhance outer space security. Observing the success with which nuclear exchanges were averted throughout the Cold War, a similar approach was considered in relation to outer space. The strategy of releasing intellectual property while guiding research was thus designed to harmonise the capacities of many nations to conduct space warfare and to protect space-based assets, thus further reducing the potential for any one of them to reasonably anticipate victory from fighting in that environment.

The key to success with many initiatives aimed at enhanced security or stability in outer space seemed likely to reside with actions by States other than the most powerful global entities. In that context, potential initiatives included: creating a forum in parallel with the UN, formulating revised and enhanced international rules for the protection of outer space, and re-evaluating entanglements with major State powers in addition to strategies mentioned already.

Activities with economic implications were recognised also. These included States distributing across power blocs and corporations their investments in space-based assets and in acquiring space-data. Support for the global economy was advocated as an additional means of encouraging global interdependence.

The initiatives here proposed and others compatible with them could become hazardous in the face of determined opposition by all major State powers. An orchestrated repudiation by them is a realistic possibility, particularly because some strategies identified could be extended to aspects of global stability additional to the security of outer space.

Ostensibly, the concept of relatively small States challenging the authority of a resolute UN Security Council would be irrational and could be portrayed as such. However, the Security Council's authority, as well as its capacity to act effectively, might be more fragile than implied by terms of the UN Charter. The Security Council's five permanent members, in aggregate, represent approximately 2.5 % of the world's total number of nation States and contain 27 % of the global population, about 19% of which is in the PRC.[1] Four of the five States are predominantly Caucasian and all five are States of the northern hemisphere. They have consistently demonstrated minimal capacity to curtail each other's activities, as was also the situation in the League of Nations during the 1920s and 30s.

In addition, Security Council concern over an alternative agenda for change might be muted in the face of obvious enthusiasm by many members of the General Assembly for amended arrangements. The additional threat of protracted pressure for changes to permanent Security Council membership, so the Council better reflected geopolitical developments of recent decades, could be disturbing and might also modify Security Council enthusiasm for opposing initiatives suggested.

Precedents such as those applying to landmines and cluster munitions have demonstrated that non-participation by major States in any given arrangement might be tolerable. A bifurcated reaction with the US, possibly supported by France and the UK, taking one position while conceivably Russia and the PRC were joined in opposition to it, could also be manageable especially if the two sides negated each other through diplomatic and related manoeuvring.

Given most or all of those developments could emerge from the advocacy of a disenchanted but energised majority of States, permanent members of the Security Council might consider it prudent to allow an emerging and parallel organisation some leeway.

Nevertheless, committed opposition from any powerful State would be regrettable, which points to a requirement for purposeful preparation of global opinion.

Several strategies with potential to be adopted globally and implemented cooperatively were therefore proposed. They included arrangements to avoid space-related false alarms, to further develop terrestrial alternatives for space-based services and declaration of a zone of non-aggression for protection of GNSS spacecraft.

Diplomacy will thus likely be an indispensable enabler of the widespread concurrence necessary to achieve progress. International conversations could be lubricated by open acknowledgement of benefits enjoyed by less powerful States as a direct consequence of powerful nations having permitted gratuitous access to space-sourced data and services.

I freely acknowledge that these are ambitious notions. Overall, the potential for success or failure of any initiative contemplated is unknowable. The latent capacity of nation States to attempt fundamental and enduring reforms leading to enhanced security of the outer space region is nevertheless apparent. Other successful geopolitical developments have originated with audacious proposals.

Contrasting this positive vision, justifiable anticipation of future and indefinite warfare is innate to the behaviour of most or all nation States. Patriotic literature, traditions, uniforms, music, medals and other symbols have combined to present warfare as an integral and sometimes revered facet of life. Those attitudes persist despite the reality of nuclear-weapon-armed ICBMs existing in numbers sufficient to destroy all extant civilisations and possibly to cause an environmentally irrecoverable nuclear winter as well.

But outer space is different. While clearly not quarantined from geopolitical ambitions, this essentially uninhabitable region experiences a diluted version of them. It is not encumbered by jingoistic recollections of past battles, neither is it fragmented by discernable national borders or polluted with visible paraphernalia of military traditions. The potential to achieve an enduring absence of warfare in outer space therefore appears to be much stronger than any corresponding ambition for planet Earth could be, unfortunately.

APPENDIX 1. IDENTIFICATION OF THE OUTER SPACE BOUNDARY

Governments of most States are acutely aware of the physical extent of their national territories situated on the land, or in some cases across adjacent seas and other bodies of water. Claims are promulgated, borders commonly patrolled or guarded, customs, quarantine and immigration facilities are established.

In contrast they pay minimal attention to the vertical limits of their legitimate sovereign authority. This appendix proposes and justifies an identifiable, vertical boundary between airspace and outer space.

Legal Basis for the Separation of Outer Space

The *Treaty on Principles Governing the Activities of States in the Exploration and Use of Outer Space, including the Moon and Other Celestial Bodies* (Outer Space Treaty (1967)) makes clear that the exploration and use of outer space, including the Moon and other celestial bodies, shall be carried out for the benefit and in the interests of all States and that there shall be freedom of scientific investigation in outer space (Article I). This is reinforced by Article II which states that outer space including the Moon is not subject to national appropriation by claims of sovereignty, use, occupation or any other means. Customary practice, supported by Articles I and II of the Outer Space Treaty, is that space vehicles may legitimately orbit over any terrestrial location while collecting data and providing other services as determined by the satellite owner or controller.

In sharp contrast, the *Convention on International Civil Aviation* (the Chicago Convention (1947)) provides for complete and exclusive State sovereignty over national airspace. The concept of sovereignty incorporates rights to repulse unwanted intruders, to allow or forbid the entry of nuclear or other weapons and to impose all of the regulations or constraints associated with national ownership.

Because there is no internationally identified geographic boundary between airspace and outer space, doubt and confusion must exist as to which of two legal regimes applies to an object travelling above the altitudes known to be achievable by aircraft or high-altitude balloons.

This uncertainty has implications extending down to sea or ground-surface level. Whether an object falling to the ground is or is not a *space* object could affect processes applicable to legal liability for damage in accordance with the *Convention on International Liability for Damage Caused by Space Objects* (Liability Convention (1972)) or lead to litigation in accordance with air navigation legislation. If a craft lands in the sovereign territory of another State and the craft is categorised as a *spacecraft* then it should be returned to the launching State as required by the *Agreement on the Rescue of Astronauts and Return of Objects Launched into Outer Space* (the Rescue Agreement (1968)). If it is not a spacecraft then domestic laws including allegations of criminal conduct such as espionage might be applied.

Regardless of methodology adopted, an enduring definition of the lowest limit to outer space will need to meet several criteria. Specifically, any definition likely to gain broad international acceptance must be: compatible with technological realities; in harmony with established science; capable of being applied using transparent and independently verifiable calculations; consistently valid across time; and applicable to any geographic location on Earth. Anything less may become immersed in avoidable controversy and will be vulnerable to reasonable repudiation.

Strategic Justification for the Separation of Outer Space

Emerging political and technological developments have been increasing the need to define a lowest limit of outer space. Four such developments are here considered: the exoatmosphere concept, the emergence of space planes, the implications of missiles and the development of hypervelocity vehicles. Singly or in combination those issues justify adoption of a clear, measurable and verifiable definition of the lowest limit to outer space.

Regarding the exoatmosphere concept, either airspace extends vertically until it intersects with the lowest limit of outer space, as implied by the Outer Space Treaty, or there exists another geographic zone in between airspace and outer space.

One such hypothetical zone, here termed the *exoatmosphere*, has been commonly inferred to exist in an alleged gap between recognised and controlled airspace and the lowest limit of outer space. The term exoatmosphere appears in official documentation, particularly in the US, where exoatmospheric might have been accepted as a credible, albeit domestic, concept.[1] The PRC, Russia, India and Europe generally are aware of the term. It has been used in news media.[2]

A State seeking to legitimatise the existence of a claimed exoatmospheric region might cite as its justification several precedents where other pre-emptive actions have become recognised as components of international law. Examples include the *Policy of the United States with Respect to the Natural Resources of the Subsoil and Seabed of the Continental Shelf* (Truman Declaration (1945)) where the US unilaterally claimed rights to its continental shelf, unilateral declarations of Air Defence Information Zones (ADIZs), or unilateral declarations of exclusion zones over the high seas for weapons testing.

Some sources gone so far as to identify specific altitudes within the alleged exoatmospheric zone, invariably without describing the zone's upper or lower boundaries. The lowest such claim I found was 48 km but other altitudes located were 61 km and 300 km.[3]

The exoatmosphere concept potentially generates at least three challenges for global stability and security.

First, broadly based international acceptance of the exoatmosphere concept could give strategic and tactical flexibility to powerful nations intent on developing and deploying weapons or weapon platforms intended to penetrate regions which otherwise, depending on altitude, might have been regarded as sovereign national territory. This could give the most militarily capable States access to regions abutting the ceiling of controlled airspace belonging to relatively smaller and weaker States.

Second, recognition by international law of an exoatmosphere would, for every State, result in the most significant loss of vertical sovereignty since 1967 when Article I of the Outer Space Treaty effectively imposed a vertical limit to sovereign airspace, without specifying an altitude.

Third, depending on the claimed upper limit of an exoatmosphere there exists the possibility some States would use it as justification for orbiting weapons of mass destruction, while claiming the weapons had not entered outer space and therefore remained in compliance with the Outer Space Treaty, Article IV, which bans them from outer space.

At present (2015) the defining spatial limits and the characteristics of an exoatmosphere are unknown. The zone has no formal identity in international law. There are no international conventions in place to describe it physically or spatially, to categorise the ownership of it, to determine permissible activity within it or to record legal or illegal acts affecting it.

I located no treaties, customary laws, or writings by eminent jurists to provide a foundation for further deliberations regarding the existence of an exoatmosphere. However, the potential for proclamation and eventual acceptance of an exoatmosphere is an emerging hazard and one which could be negated by declaration of an identifiable lowest boundary to outer space.

The concept of vehicles able to both fly and orbit thus functioning in both the terrestrial environment and the outer space region is well-established. To date, rocket-propelled vehicles have demonstrated a capacity to launch from Earth into outer space, perform manoeuvres in outer space, including docking with another space vehicle such as the International Space Station, before returning to Earth and being landed either by a human pilot or autonomously. However, more dramatic claims and complaints regarding their capabilities have been both made and inferred.[4] The evolution of space plane technology was considered in Chapter 5 under *Space Planes and Hypervelocity Vehicles.*

The PRC has described US space plane developments as '*widely regarded as the next generation super weapon that is even more dangerous than* [the] *atomic bomb*.'[5] The Russian news agency, *Pravda*, published an article titled, *Secret American Space Planes to Dominate Planet Earth*.[6] Iran has also been concerned by US developments in relation to space planes.[7]

The possibility exists of such vehicles claiming to be *aircraft* and carrying weapons of mass destruction or for their crews to seek special immunities from charges of espionage on the grounds they are *spacecraft*. This strategic flexibility, associated with the absence of a defined lower limit to outer space would be contrary to the interests of almost all States. A major spacefaring State might also rue these hypothetical developments if it became the recipient and not the initiator of aggressive or intrusive activities conducted within a cloud of jurisprudential imprecision.

The challenge of regulating vehicles able to function in more than one environment is neither new nor particularly complicated. Numerous other vehicles have successfully functioned in more than one environment and have been required to comply with the regulations governing their current environment at any given time. Seaplanes and floatplanes for example are typically liable to follow aeronautical procedures when airborne and maritime rules when on the surface of a body of water. Amphibious automobiles are required to obey the rules of the road when on land and maritime rules on water. In the case of a space plane, promulgation of specific rules describing acceptable launch and landing trajectories within the dense atmosphere could be undertaken as is routinely done for aircraft.

The legal status of space planes or similar vehicles would be clarified, the risk of controversial attempts to destroy them minimised, and their capacity to intrude into the territory of sovereign States reduced if the air environment and space region were formally delineated. The fact some vehicles may be capable of functioning in two or more physical environments is not a justification for asserting they should be exempt from categorisation or related regulation. Any exemption would run against well-established national precedents.

Since the first flying bombs struck London in January 1944 nations have attempted to deploy defences against missiles travelling toward or over their territory. The US installed anti-missiles defences at the earliest time technologically possible as did the then USSR.[8] The US is currently pursuing a missile defence program known as National Missile Defense.[9] Russia also has increased investment in anti-missile technologies, and several other States have developed or purchased defences against attacks by missiles. Russia and the US are among States which have exported systems to defend against missile attack.

Contemporary States seem to assume that the flight of a missile across their territories would be an intrusion and potentially one inviting a hostile response.[10] This approach apparently reflects a historical understanding that State airspace or sovereignty extends vertically for an unlimited distance. The unlimited distance concept was implicit in the *Convention for the Regulation of Aerial Navigation* (Paris Convention (1919)), Article 1, which gave each signatory '*...complete and exclusive sovereignty of the airspace above its territory*.' This component of the Paris Convention was reiterated in the 1947 Chicago Convention before being modified fundamentally by the Outer Space Treaty of 1967.

After 1967 attempts to engage any alleged missile, merely because it was passing over, but not necessarily within, sovereign territory could lead to a sustainable allegation of aggression against an object legitimately traversing the global commons of outer space. One response to this assertion, based on the notion of self-defence, might be that the capacity to characterise a missile as potentially hostile was available.[11] This, however, assumes detection of the launch and observation of the missile's trajectory combined with a capacity to estimate a point of impact – an almost futile process when modern missiles could be reprogrammed in mid-flight.

One relevant incident occurred on 5 April 2009 when the DPRK launched a Taepodong-2 Missile, allegedly across part of Japanese territory. Being aware in advance of DPRK intentions to launch, Japan had threatened to deploy missile interceptors.[12] US President Obama was reported as calling for a global response.[13] The incident was reviewed by the UN Security Council which, although apparently united in condemnation of the DPRK, was unable to agree on an appropriate reaction.

Records indicate the DPRK had remained fully aware of the jurisprudential flexibility available to it and could claim to have acted legitimately regardless of Security Council endorsement or opposition.[14]

On 12 December 2012 the DPRK used the launch vehicle Unha-3 to launch its *Kwangmyongsong-3* satellite into orbit. The UN Security Council again responded with a resolution condemning the DPRK's action.[15] In this case also any claimed legal capacity for Security Council or national intervention, armed or otherwise, could have been clarified had the lower boundary to outer space been defined, or rendered more obscure if the exoatmospheric notion had been injected into the debate.[16]

Hypervelocity vehicles, considered in Chapter 5 at *Space Planes and Hypervelocity Vehicles*, represent an emerging technology related to missiles. The high-velocity and high altitude operations involved suggest, if they are eventually developed to carry offensive payloads, these vehicles could operate within the vaguely conceptualised exoatmospheric zone. These realistic possibilities provide one more incentive for technologically advanced States to oppose adoption of a defined lowest boundary for outer space and yet more justification for the majority of other States to insist on clarifying their vertical sovereignty, thus delineating the territory they can defend and control.

Attempts to Identify the Lowest Limit of Outer Space

Many attempts have been made to define or describe the lowest limit to outer space. I offer only four here as examples:

a. H Bhayana, 2001, claimed, '*In the broadest sense, space may be regarded as covering those reaches of the universe which are beyond the earth's sensible atmosphere.*'[17]

b. The von Karman line allegedly asserts that outer space starts at 100 km apparently without further elaboration. It has been used as a discriminator by Fédération Aéronautique Internationale to define the start of outer space.[18]

c. Australia's Space Activities Act, 1998, approached but then avoided a committed definition of the lowest limit of outer space. Section 8 of the Act defined a launch into space as ' ... launch the object into an area beyond the distance of 100 km above mean sea level, or attempt to do so.' The unstated implication was that the lowest limit of outer space began immediately beyond the 100 km level.19

d. The US adopted a domestic law effectively defining the start of outer space as being at an altitude of 50 miles (approximately 80 km) because attainment of that altitude became the altitude-related prerequisite for award of an astronaut insignia.[20]

The Bhayana concept is logically circular in nature and effectively substitutes the term '*beyond Earth's... atmosphere*' for that of outer space while adding a further requirement to quantify '*sensible*' as the defining point of demarcation. The remaining three definitions, which employed altitude as the primary discriminator separating the terrestrial and outer space regions, offered no overt explanation or justification for adopting their chosen altitude. Neither did they discriminate between pressure and linear altitude. Nevertheless, altitude-dependent definitions are nominally quantifiable, rational and easy to comprehend. They have merit, at least in principle.

Physical Observations of the Near-Earth Environment

No study I located identified any environmental phenomenon forming a consistent boundary between the airspace environment and outer space region. I have employed only two examples, atmosphere and gravity, to illustrate the challenges inherent in attempting to do so, although similar analyses could be undertaken with regard to many other physical properties.

Earth's atmosphere exists in generalised layers with transition boundaries between them as shown on the simplified diagram, Figure A1, *Atmospheric and Thermal Profile Above Earth*. But the relationship between altitudes and atmospheric bands shown in Figure A1 is merely indicative; first because published references vary in their estimates of the dimensions of each atmospheric layer and secondly because the vertical dimensions of the layers are dynamic in relation to both time and to geographic location. The diagram omits reference to the ionosphere, to regions of Van Allen radiation, to the magnetosphere, to the sodium and ion belts and other phenomena of the upper atmosphere, but detail provided is adequate to support the critical point made next.

Figure A1. Atmospheric and Thermal Profile Above Earth [21]

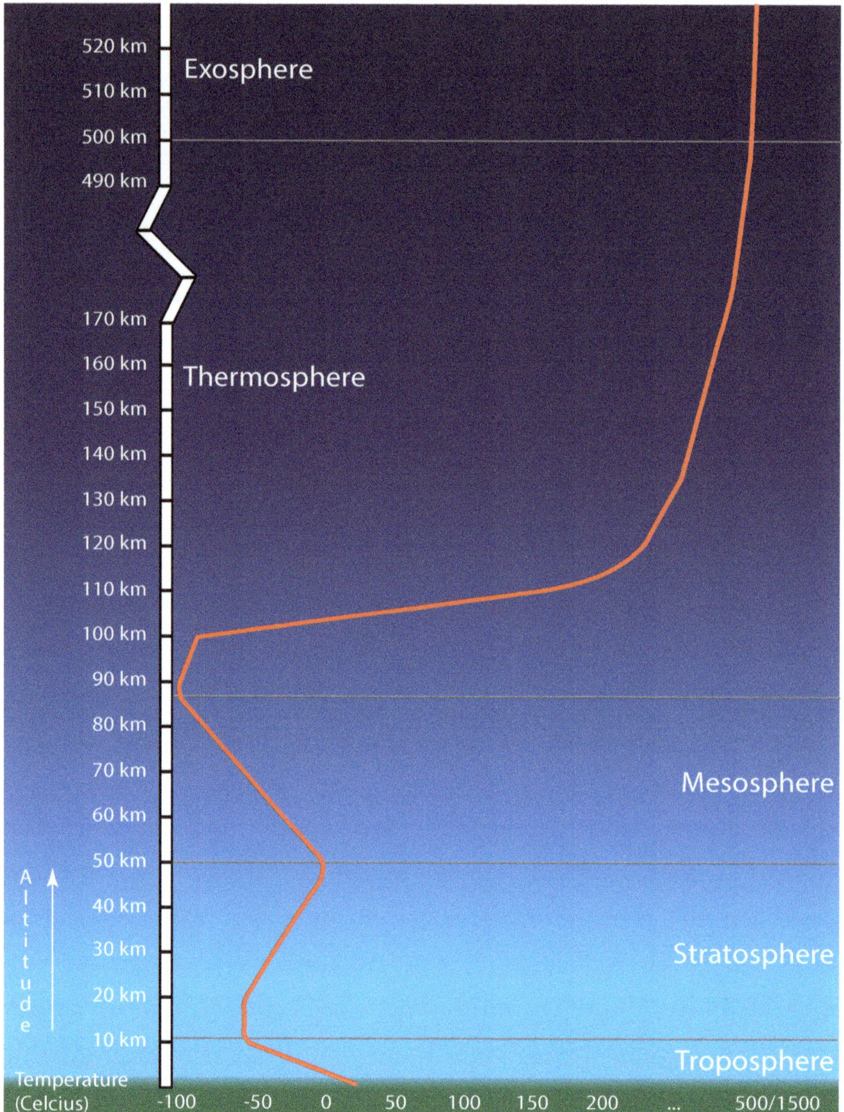

Atmospheric bands associated with Earth clearly extend to altitudes above those at which satellites and other spacecraft can and do remain in stable orbits, with many of them orbiting in the altitude band 800 – 1 000 km and some lower than that. In particular the exosphere (as contrasted with the fictitious 'exoatmosphere') a region populated primarily by hydrogen and helium atoms, together with representations of other gases near to its base, continues to at least 10 000 km above Earth although the diagram shown ends at ~520 km.

This is a potent observation because the majority of spacecraft ever placed into Earth-centric orbit have been and remain at altitudes less than 10 000 km, regardless of how that altitude is defined or measured.[22] If, therefore, transit through an identifiable component of Earth's atmosphere meant that a vehicle was not in *outer space* then many satellites would be orbiting in national sovereign territory and liable to interdiction.

Neither can an identifiable measure of gravity indicate a boundary between airspace and outer space. The force of gravity changes constantly in relation to spacecraft in orbit around Earth. The strength of the gravitational force exerted by planet Earth varies between terrestrial locations and also, to a lesser extent, as a function of time. Spacecraft are additionally affected by extraterrestrial sources of gravity, including gravity from the Moon and the Sun. Depending upon their locations relative to Earth, gravity exerted on spacecraft by the Moon and the Sun is at times cumulative, at other times competitive. On occasions their gravitational attractions operate tangentially in relation to Earth-centric orbiting objects.

These examples typify the attributes of natural phenomena surrounding Earth. The natural phenomena vary as to altitude, density, intensity, seasonality, position in relation to terrestrial latitudes and in some cases all of those. In summary, there exist no observable physical phenomena which could, singly or in any combination, be employed to consistently identify or demarcate a chronologically and geographically persistent outer space region.

An additional complication when trying to define the start of outer space is that planet Earth is flattened at its poles and bulges at its equator. It thus is not a perfect sphere. The equatorial radius of Earth is approximately 6 378 km whereas its polar radius is approximately 6 357 km, a difference of some 21 km caused by a flattening at the poles with a ratio of approximately 1 in 304 in relation to the equatorial radius. The preferred mathematical model of Earth's shape is an oblate spheroid, or reference ellipsoid. An extrapolation of this profile vertically into outer space would create a secondary boundary which would mirror the planetary geoid and so would also not be a perfect sphere.

Taking account of these limitations and complications, a reliable and consistent description of the interval between the surface of Earth and the lowest boundary of outer space will require a verifiable datum from which to take appropriate measurements. The datum will of necessity be neither simple nor intuitive.

The Datum

In addition to the challenge of Earth's surface profile not being a perfect sphere, the planetary surface is affected by observably dynamic influences. Tectonic plates, which include regions above and below sea levels, are mobile.[23] Every total oceanic tidal cycle on Earth takes 19 years to complete.[24] Other terrestrial influences include diurnal and seasonal atmospheric pressure changes and cumulative sea level increases caused mainly by melting of polar ice caps.

Geodetic science can surmount all of these challenges although the techniques employed are at times complicated and therefore difficult to communicate and to apply in practice, particularly when measurements involving extragalactic observations are employed, as commonly happens.

It is nevertheless possible to use an established geodetic criterion describing a defined global baseline; for example the criterion based on the distribution of the mass of the solid Earth. In this context Earth Gravitational Model 2008 (EGM 2008) represents a comprehensive current model of the Earth's equipotential surface that corresponds to the mean ocean surface. In effect it describes the consequences of a tide-free and current-free ocean, via the geoid surface, in a manner following hypothetical pathways cut through all continents and other land masses.

Being a gravitational model EGM 2008 incorporates gravitational variations due to variations in Earth's mass distribution, and these result in vertical variations with respect to the best fitting reference ellipsoid, ranging from –107 m to +86 m over different locations on the Earth's surface. By agreement, when determining a datum for defining the lowest limit of outer space, these variations could be either incorporated within a datum to support the lower boundary definition, or a uniform adjustment might be applied to simplify the complicated surface. However, any simplification would introduce errors relative to the datum, in the order of up to approximately 100 m, and could reduce the traceability of the datum.

The computational methodology supporting EGM 2008 is transparent and independently verifiable. It is reviewed and strengthened periodically. The datum is enduring. (The previous version was EGM 1996). It is traceable, and is the best approximation available or likely to become available within the foreseeable future to the physical gravitational field of the Earth. The EGM concept is well known and is embedded in numerous geodetic models. EGM 2008 is globally accepted by the geodetic community.

However it is undeniably complicated, even in a simplified format with variations removed or uniformly resolved statistically. Fortunately some of the computational complexity may be of minimal consequence given the ready availability and accessibility of computing power and the transparency of underlying formulae. It is shown at Figure A2, *EGM 2008 2.5 Minute Geoid Heights.*

Figure A2. EGM 2008, 2.5 Minute Geoid Heights. [25]

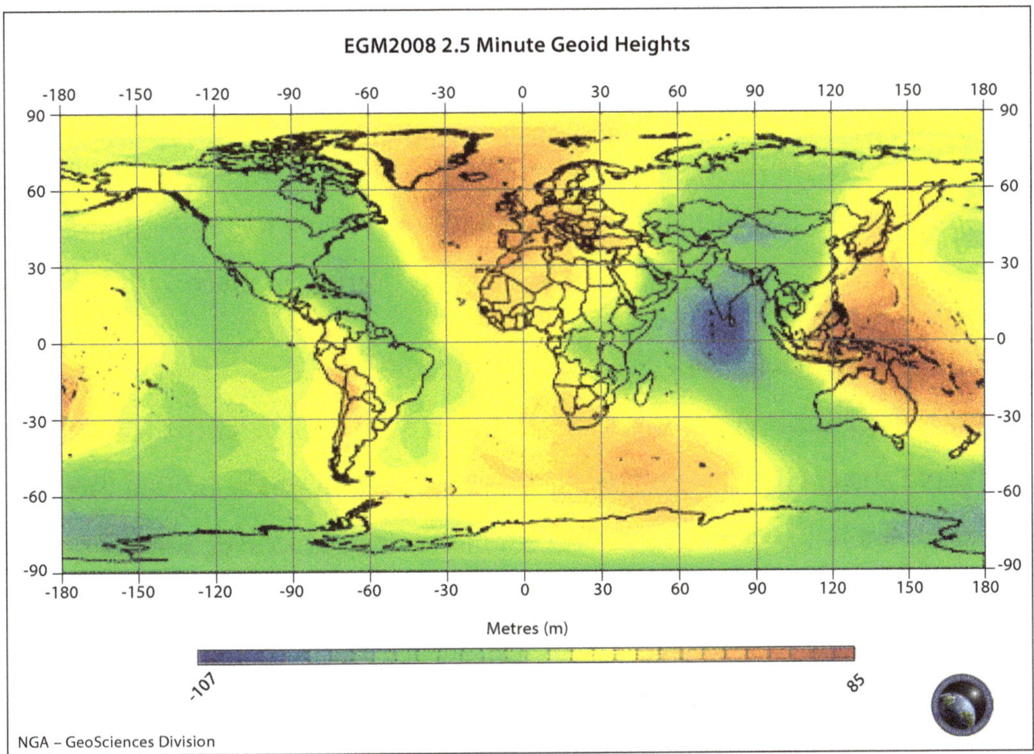

EGM 2008 is accurate to within several decimetres in its vertical definition of the geoid. Geodetic techniques, in particular those applied to detailed regional surveys, can achieve yet higher accuracy. However given the scale of measurement involved when determining a lowest boundary to outer space, additional accuracy would seem superfluous and therefore unnecessary. Additional accuracy would almost invariably be accompanied by requirements for more frequent adjustments to reflect subtle variations in the gravitational distribution thus introducing an element of continuing and arguably unjustifiable change.

EGM 2008 could offer a uniform and global datum from which every State could measure to the upper limit to its claimed sovereign territory and comprehend where the global commons of outer space began over its territory.

Perpendicular Measurement

The second and indispensable step in determining a lowest limit to outer space is to apply a perpendicular and linear altitude measurement beginning at any reference point upon the accepted datum surface. Use of linear measurement is desirable because, although definitions of altitude based on atmospheric pressure exist, the associated physical distances are inconsistent over time.

The first step in this further process is to identify upper and lower limits to the altitude range from which this defining altitude above Earth should be selected. It will be critical to ensure any delineating altitude selected is above the height attainable by any conventional form of aviation and below the height at which any spacecraft could sustain a ballistic orbit.

It might be argued there is no such altitude because evolving space planes and related engines will be able to operate in or transition across a significant range of altitudes. That claim, however, ignores the reality of extant legal and regulatory regimes applying effectively to numerous other multi-environment vehicles such as floatplanes and amphibious automobiles.

The lowest altitude which might be considered as defining the beginning of the outer space region will be guided by the upper limit of aviation. In this context, the term *aviation* is proposed to include all aircraft or balloons, the vertical sustainment of which is related to ambient atmospheric gases.

Excluding rocket-powered craft the current (2015) aircraft altitude record appears to be 29. 524 km set by a propeller-driven and solar-powered aircraft in 2002.[26] The maximum altitude achieved by a balloon is 53.7 km (in 2013) although attempts will be made to achieve higher altitude records.[27]

Based on this evidence it seems reasonable to accept that the atmospheric environment required by vehicles needing interaction with ambient gases to maintain altitude will extend to a linear maximum of less than 60 km. Thus 60 km may be taken as establishing a ceiling for conventional forms of aviation. It can be accepted as the lower limit for the altitude range within which outer space might be considered to begin.

The higher limit for the range of altitudes in which outer space may be regarded as starting, should be the lowest altitude at which an artificial satellite could sustain a ballistic orbit. It then becomes necessary to consider how much travel unassisted by artificial propulsion is required to make an 'orbit'. Would an arc comprising one quarter of one orbit suffice, or would several completed orbits be required; if the latter then how many completed orbits would be needed and who has authority to make such a determination?

In reality the minimum linear altitude at which an orbit can occur is and will remain unknown. It will be influenced not only by definitional challenges but also by many dynamic natural factors. The United States' KH 7 satellites have been reported as occasionally declining to a perigee of 120 km.[28] A document released in 2011 by the US National Reconnaissance Office has recorded a satellite being subjected to seven revolutions at 113 km with no apparent spacecraft problems.[29]

However, so far as I could determine no spacecraft had ever maintained a ballistic orbit at a linear altitude less than 100 km. In the absence of any conflicting report or data it therefore seems reasonable to accept this as the lowest altitude at which a spacecraft could sustain an orbit. It is therefore maximum altitude in the altitude range from which the start of outer space should be selected.

From the foregoing deliberations the altitude least likely to ever involve a conflict with human operations in either airspace or outer space would be at the middle of the 60 – 100 km range; namely the 80 km level. Regardless of any debates concerning the appropriateness of the 60 or 100 km criteria, on contemporary evidence no vehicle requiring atmospheric interaction to maintain altitude will be able to operate at 80 km and no space vehicle will be able to maintain a ballistic orbit at that altitude either.

The 80 km concept is almost identical to the definition used by the US in its domestic legislation (50 miles), already mentioned, pointing to either an astonishing coincidence or to an indication the US has already made and accepted calculations similar or identical to those above.

In summary, the proposed and justifiable definition of a lowest boundary to outer space is that outer space begin at a linear and vertical distance of 80 kilometres above the datum EGM 2008. This might by agreement be updated periodically to reflect globally endorsed refinements to the EGM datum.

Organisational Mechanisms

Despite increased requirements to adopt an internationally recognised lowest limit to outer space, and the enhanced technological capacity to do so, little or no progress has been made toward achieving that objective within international forums.

Under the terms of *The Charter of the United Nations*, members of the UN General Assembly could in theory agree to specify the boundary to outer space in terms applying to the sovereign territory of assenting members as well as to regions over high seas and the Antarctic. Any State Member of the UN General Assembly could raise this issue. In particular, the UN Charter, Article 11 allows the General Assembly to consider general principles of co-operation in the maintenance of peace and security.

However, if the issue of clarifying the boundary to outer space was raised in the General Assembly, history points strongly to the General Assembly following its practice of more than 50 years and referring the matter to specialised agencies such as the Committee on the Peaceful Uses of Outer Space (COPUOS) or possibly to the International Civil Aviation Organisation (ICAO) or to both.

On 12 December 1959, the UN General Assembly made COPUOS a permanent committee with membership of 24 States. By 1980 COPUOS membership had reached 53 States and by 2010 its membership totalled 70 States. It therefore appears to be an organisation which should be able to pronounce authoritatively on issues concerning outer space.

Regarding the need to identify a lowest limit of outer space, it was mentioned in a 1959 report by the then *Ad Hoc* COPUOS but no such definition was contained in the subsequent Outer Space Treaty. In 1966, when the UN General Assembly agreed to the resolution commending the Outer Space Treaty for signature, COPUOS was requested to begin to study the definition of Outer Space.[30] The issue was placed on the agenda of the COPUOS Legal Subcommittee, but no discernible progress was achieved during the following ten years.[31] The *title* of the request was then revised.

In its 2006 report to COPUOS the Legal Subcommittee, still pursuing the definition of the lowest limit of outer space, decided to invite participation by the International Telecommunications Union. I did not find the subsequent contribution of the International Telecommunications Union, to identifying the lowest limit of outer space.

Records of the 2007 proceedings of the Legal Subcommittee reflected a well-established divergence of States' opinions on defining outer space. One cohort considered the delimitation of outer space to be important for determining the scope of applications of air law and space law. Another held to the view that States should continue to operate under the current framework, which was claimed to be functioning well, until such time as there was a demonstrated need and a practical basis for developing a definition and delimitation of outer space.

Despite this spread of perceptions three initiatives relating to the definition of the lowest limit of outer space were adopted. These were:

a. *'The view was expressed that the tendency of using the lowest satellite orbit as a criterion for the delimitation of airspace and the lowest limit of outer space was obsolete in view of the fact that both the X-15 rocket plane and SpaceShipOne were regarded as spacecraft and qualified as suborbital, which, according to the criterion, meant that the beginning of outer space could be far below the lowest satellite orbit.'*
b. *'The view was expressed that progress in the definition and delimitation of outer space could be achieved through cooperation with the International Civil Aviation Organisation.'*
c. The Legal Subcommittee convened a Working Group on the Definition and Delimitation of Outer Space.

The Legal Subcommittee's 2008 report to COPUOS included a new (for COPUOS) approach to defining the lowest limit of outer space. It proposed that *'if member States failed to give clear-cut criteria for délimitation [of outer space], a special regime or zone between airspace and outer space should be explored.'* The rationale for this proposal, impliedly a concept supporting the exoatmosphere concept, was not made clear.

In its 2008 report, COPUOS recorded a report from its Working Group on the Definition and Delimitation of Outer Space. The Working Group had achieved little. Its report recorded a decision to invite member States to consider if a definition of the lowest limit of outer space was necessary or whether such member States could advance alternative proposals.

The Legal Subcommittee's 2009 report to COPUOS was notable only in that some delegations were recorded as having complained about little progress being made toward defining the lowest limit of outer space despite it having been on the Subcommittee's agenda for more than 40 years.[32] The 2010 report to COPUOS did not record any progress toward defining the lowest limit of outer space.

Records of attempts within COPUOS to define the lowest boundary of outer space suggest the Committee will not succeed in agreeing to a definition. The only realistic prospect of such a definition emerging would be for COPUOS to refer the matter back to the General Assembly. Even this might not be achievable. Because COPUOS apparently functions on a consensus basis, the cohort opposing identification of a definition of outer space might be able successfully oppose a reference back to the General Assembly.

Notwithstanding the apparent paralysis in COPUOS, a declared *upper limit to international airspace* could act as a surrogate for the lower limit of outer space. It could dilute or even negate the need to define outer space. It would also effectively remove the notion of an exoatmosphere. This, conceivably, was the concept which motivated COPUOS in 2006 to propose referring the boundary definition issue to the International Civil Aviation Organisation (ICAO). By 31 October 2013 the list of contracting States to the ICAO numbered 191 and so it was approaching duplication of membership of the UN General Assembly.[33]

Some components of the Chicago Convention appear to support the notion that ICAO might have capacity to recognise an absolute upper limit to airspace. This would leave States members of the organisation implementing that outcome through their domestic arrangements. Chapter 1, Article 1 of the Convention declares all States have complete and exclusive authority over airspace above their territories. It may therefore be reasonable to conclude that resident within ICAO is the authority able to define the limit to that airspace. Chapter 2, Article 12 of the Convention authorises the establishment of rules for flight over the high seas and this could, in conjunction with domestic legislation, provide widespread coverage of a ceiling to airspace.[34]

However, other components of the Chicago Convention do not support that impression. The preamble to the Convention, which is not necessarily part of the Convention itself, identifies one purpose of the agreement as being to develop international civil aviation in a safe and orderly manner, an objective not associated with defining a surrogate limit to outer space. Chapter 1, Article 3 a, states that the Convention applies only to civil aircraft and not to State [military and related] aircraft. Any determination made in the context of the Convention could therefore have only limited application in this regard and might exclude the high altitude vehicles most likely to be engaged in hostile activity. Through Chapter 1, Article 4, of the Convention, every party agrees to not use aviation contrary to the purposes of the Convention. In this context, it is questionable whether using civil aviation regulatory mechanisms to produce a surrogate lower limit to outer space could be an acceptable application of the Convention. A surrogate lower limit to outer space would, by definition, be an altitude at which no machine requiring support from the air could conceivably fly, and it is therefore questionable whether an organisation such as ICAO, with its focus on regulating aircraft or air travel, has the legal capacity to make any determination in that context.

An additional limitation to employing ICAO, as the forum through which a surrogate limit to outer space might be defined, is that the space boundary problem is merely one of numerous and related challenges confronting the development of international law regarding space. Other issues are considered in Chapter 6. *International Law and Outer Space.* The distribution of these issues across carriages of administrative convenience, even if legally and administratively possible, could entail risks of subsequent obfuscation of space law and a reduction in capacity to coordinate the international law of space.

In summary, progress toward a legal definition of the lowest boundary of outer space has not been achieved within COPUOS. Given the practice within COPUOS of decision making by consensus, future progress by COPUOS toward achieving an agreed definition is highly unlikely while some States members oppose that outcome. There are also grounds to doubt that ICAO has legal capacity to determine a surrogate lowest limit to outer space. Argumentation over the issue of legal capacity would likely stall progress within ICAO indefinitely.

It is difficult to determine the relevance of the International Telecommunications Union (ITU) to identification of a lowest limit to outer space, although that organisation was mentioned in the 2006 COPUOS report. ITU functions are focussed upon radio communications and frequency management.

Arguments Against Defining the Lowest Boundary to Outer Space

Given the recorded inability within COPUOS to agree on a definition to the lowest boundary of outer space, objections in any alternative forum are likely. Following is an examination of but three predictable submissions opposing a definition of outer space.[35]

The claim has been made that requirements to define the boundary to outer space have not arisen. Two responses are readily available. The first is, given doubts regarding the boundary of outer space, some or many unauthorised intrusions into prospective sovereign airspace might already have occurred but they could not be clearly categorised. Neither for the same reason could they be convincingly denied. With the development of space planes, hypervelocity vehicles, the problem of missile overflight and the notion of an exoatomosphere, the need for delineation has been increasing. This would point to a strengthening requirement for adopting a definition and not to a reason for avoiding one.

Secondly, a definition of the lowest limit of outer space would function as a pre-emptive precaution against intrusive and controversial encroachments into sovereign territory. Pre-emptive strategies are neither new nor unique, they having been adopted by many States. Those States include France and Britain with their invasion of Egypt in 1956, the PRC which invaded Tibet in 1950, the US which led an invasion of Afghanistan in 2001 with Britain as a major ally, and by the USSR which invaded Afghanistan in 1979. In summary, all five permanent members of the UN Security Council have justified and employed pre-emptive strategies when it suited their national ambitions or requirements. It would be inconsistent for States which have held, and continue to hold, policies supporting the principle of pre-emption, to oppose defining the lowest limit of outer space because it might be pre-emptive in nature.

Another predictable objection is that a definition of the lowest limit to outer space would be arbitrary. Undeniably, any altitude finally adopted as the discriminator between airspace and outer space would be arbitrary but this does not justify avoiding any such determination. There are many established examples within international law for adopting arbitrary boundaries. Part II, Section 2, Article 9 of the Law of the Sea Convention '*If a river flows directly into the sea, the baseline shall be a straight line across the mouth of the river between points on the low-water line of its banks.*' is an example of an arbitrary boundary, as is the declared 12 nautical mile limit to territorial sea recorded at Part II, Section 2, Article 3 of that Convention. Such concepts are commonly accepted and unavoidable components of international and of many domestic laws.

Arbitrary is not a synonym for capricious.

Implications

The recognition of physical boundaries, including the existing and well-established horizontal terrestrial borders of nation States, does not guarantee immunity from trespass or invasion. It does however describe geographic boundaries able under international law to be defended, it defines the region in which sovereign authority may be exercised and it provides for intrusive actions to be defined as illegal. Those fundamental components of State security do not now exist in relation to vertical sovereignty, except conceivably in a limited and strategically hazardous manner with ICAO having identified an upper limit of *controlled* airspace and that definition having commonly been adopted domestically.

Foreseeably, an attempt to define the lowest limit of outer space will contain some imperfections requiring subsequent clarification. This reality has not prevented the evolution and development of international law in many other respects and it does not provide a substantive argument justifying inaction in this situation.

An identifiable boundary distinguishing airspace from outer space is required. The approach advocated here is: compatible with technological reality, in harmony with established science, capable of being ascertained using transparent and independently verifiable calculations, consistently valid across time, and transferable to any geographic location on Earth. The degree of accuracy obtainable would suffice for applications involving a definition of the lowest limit to outer space.

The achievement of a broadly accepted definition of the lowest boundary to outer space is likely to require action within a forum functioning parallel to the UN, its committees and its agencies. Alternatively, it is within the capacity of individual nation States, possibly with several acting cooperatively, to promulgate definitions of the limits to their vertical sovereignty.

Appendix 2. Nuclear-Detonation-Induced Radiation Effects

> On 9 July 1962 the US in a test known as Starfish Prime detonated a 1.4 Mt bomb at an altitude of 400 km. No spacecraft was publically recorded as having been destroyed by the blast effect, but consequent radiation damaged up to six spacecraft.[1] One damaged satellite, *Arial 1,* was 7 400 km distant at the time of detonation.[2]

Most of the literature describing the consequences of detonating nuclear weapons in outer space refers to an electromagnetic pulse (EMP). However, not all electromagnetic effects caused by nuclear weapons detonated in outer space can be usefully described as a pulse. Consequently the term nuclear-detonation-induced radiation effects (NDRE) has been adopted here.[3]

Some of the effects associated with detonations of nuclear weapons can and do occur naturally. They can harm spacecraft, terrestrial power grids and electronic equipment.[4] However, information following is biased toward the capacity of anthropogenically induced NDRE to damage spacecraft. The terrestrial effects are not considered.

The primary components of NDRE are shown in the following diagram.

Figure B1. Radiation from a Nuclear Explosion. [5]

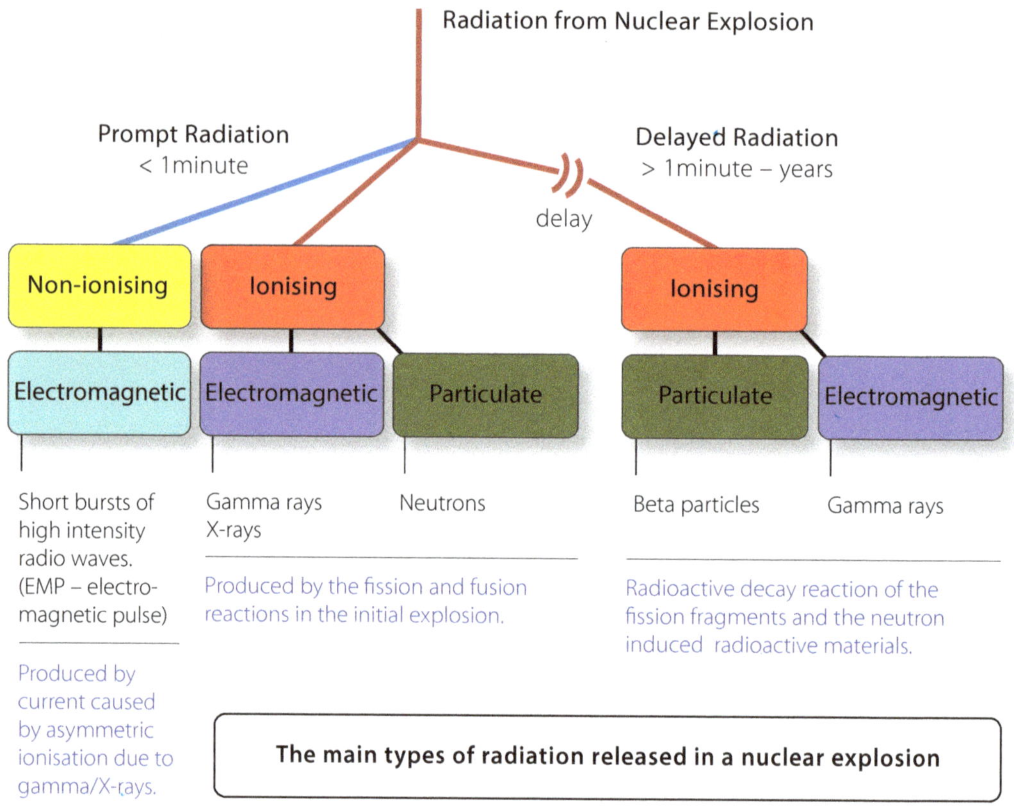

From the perspective of damage to spacecraft, a fundamental categorisation is the separation of radiation into 'prompt' and 'delayed', with prompt effects including both non-ionising and ionising radiation.

Prompt radiation, lasting between less than one second to less than one minute, can transmit energy to spacecraft skins or sheaths, including energy at X-ray frequencies. Consequent asymmetric electric currents induced by gamma radiation can enter spacecraft apertures to debilitate or destroy electronic components. Some spacecraft sheaths can be penetrated with electrons being deposited on interior surfaces leading to a capacity to produce spurious currents able to destroy electronic components.[6]

Delayed radiation incorporates components which persist for periods of from one minute up to more than a year. The radiation is primarily beta particles resulting from radioactive decay of fission products or possibly neutron-induced radioactivity from the explosion. Electrons, spiralling back and forth along Earth's magnetic-field lines, produce an enduring radiation pumping effect. [7]

Satellites encountering the pumped environment will, unless specifically designed to withstand such categories of radiation, absorb cumulative amounts until their electronic circuits degrade and eventually fail. At best, the expected effective life of these spacecraft is likely to be reduced.[8]

Techniques to protect against delayed radiation effects include incorporation of gallium arsenide (GaAs) transistors but comprehensive insulation from the long-term effects of a nuclear detonation in outer space is challenging.[9] In addition to potentially harming spacecraft, an artificially pumped radiation environment could, depending on communication frequencies and techniques used, make it more difficult for surviving higher-altitude spacecraft to communicate with ground stations.[10] This could result in the incremental loss or degradation of some space-based services.[11] Replacement of spacecraft damaged by, or with their performance degraded in consequence of, NDRE may be impracticable until the radiation effects had reduced or appropriately radiation hardened replacements became available.

Giving contemporary context to this issue, radiation hardening of high-value spacecraft to resist an intense burst of prompt electromagnetic NDRE has been regarded as technically feasible, albeit economically unappealing. [12] Many spacecraft are radiation hardened to survive only the full spectrum of radiation effects expected to occur naturally.[13] Notwithstanding precautions taken, a study in 2001 by the US Defense Threat Reduction Agency concluded:

> *'One low-yield (10 – 20 Kt), high-altitude (125 – 300 km) nuclear explosion could disable – in weeks to months – all LEO satellites not specifically hardened to withstand radiation generated by that explosion.'* [14]

Given the trend toward States utilising commercial spacecraft, the detonation of nuclear devices in outer space would impact both national and commercial operations. In 2008 The [United States] Commission to Assess United States National Security Space Management and Organisation determined that:

> *'Commercial satellites support many significant services for the Federal Government including communications, remote sensing, weather forecasting and imaging. The national security and homeland security communities use commercial satellites for critical services, including direct and back-up communications, emergency response services, and continuity of operations during emergencies ... Satellites in low Earth orbit generally are at risk of lifetime degradation or failure from collateral radiation effects...a nuclear detonation at high altitude produces numerous other effects that can impact on the performance and survival of satellites.'* [15]

Confirming its destructive potential, NDRE has on occasions been viewed as a potential outer space weapon *per se*.[16] NDRE in this role would be more attractive to States with below average reliance on data and services provided from outer space. Other space-dependent States facing a last-resort response to a major missile attack might also use nuclear weapons in a defensive role and accept the sacrifice or degradation of space assets as a necessary collateral cost associated with preserving terrestrial infrastructures and protecting human populations. Another potential motivation for detonating a nuclear weapon in outer space could be to provide an unmistakable warning of serious intent or commitment in a pending conflict.

From evidence available there is no doubt a nuclear weapon detonated in outer space would, given the contemporary population of orbiting spacecraft, cause major damage. The harm would over time be agnostic as to State of spacecraft ownership. International concern about the dangers of NDRE, more commonly referred to as EMP, is justified.

Notes

Introduction

1 EC Dolman, *Astropolitick: classical geopolitics in the space age*, (Cass series. Strategy and History; no. 4) FRANK CASS, London, 2002.

Chapter 1

1 A. Shternfeld, *Soviet Space Science*, trans. United States Air Force, Hutchinson (Basic Books), London, 1959, pp. 291 – 336. (Given governance arrangements in the USSR at that time, the book was accepted as having been State sanctioned. Author)

2 *The Pioneers, Clément Ader*, 1998, Monash University, Clayton, viewed 9 May 2012, <http://www.monash.edu.au>. (Other possible contenders for the title 'first' in this context were Felix du Temple de la Croix in 1874 and Alexander Mozhaiski in 1884.)

3 *China seen fielding nuke-capable ballistic missile*, 18 July 2011, Global Security Newswire, n.p., viewed 30 July 2011, <http://gsn.nti.org>.

4 The 90-minute (rounded) estimate was included to illustrate the compressed timeframe in which strategic weapons could be applied to targets globally. The estimate is based on a launch from Vostochny, (Russia, 51°42'N, 128°O.W) into a south-westerly trajectory which would cross the PRC and the Antarctic Continent, turning north to pass over the western States of the US, western Canada and Alaska before releasing multiple independently-targeted re-entry vehicles (MIRVs) for a back-door attack against early-warning radars operated by North American Aerospace Defense Command. With the spacecraft (AKA missile) having been launched into a 1 000 km altitude orbit and thus travelling at an average of 7.4 km/s, the time in orbit would be approximately 82.5 minutes plus a total allowance of 4.0 minutes to achieve orbit and for deceleration of the MIRVs during re-entry prior to the attack. Calculations were extrapolated from an Earth diameter of 12 742 km.

5 JV Vacca, *Satellite encryption*, Academic Press, Orlando, 1999.

6 DJ Hamer, *Bombers versus battleships*, Allen & Unwin, St Leonards, 1998, p. 2.

7 *US missile hits toxic satellite*, 21 February 2008, BBC, London, viewed 9 May 2012, <http://www.bbc.co.uk>. *US Navy's upgraded Aegis BMD system passes functional assessment,* 27 August 2010, Space News, Imaginova Corporation, Stamford, viewed 6 March 2011, <http://www.spacenews.com>.

8 *China comes clean over shot-down satellite,* 23 January 2007, New Scientist, Sutton, viewed 20 January 2014, <http://wwww.newscientist.com>. *China confirms satellite downed,* 23 January 2007, BBC News, London, viewed 24 February 2014, <http://www.bbc.co.uk>.

9 C Schumacher, *UAV cooperative control – motivation and introduction*, 2007, University of Michigan, Ann Arbor, viewed 29 August 2012, <http://www.umich.edu>. J How, E King & Y Kowat, *Flight demonstrations of cooperative control for UAV teams*, 2004, Massachusetts Institute of Technology, Boston, viewed 29 August 2012, <http://hohmann.mit.edu>. M Daly (ed.), *Jane's unmanned aerial vehicles and targets*, Jane's Information Group, Surrey, 2008.

10 *Air Force priorities for a new strategy with constrained budgets*, February 2012, United States Air Force, n.p., viewed 13 March 2013, <http://www.af.mil>. *Aging array of American aircraft attracting attention*, 27 September 2011, Defense Industry Daily, n.p., (US), viewed 29 April 2012, <http://www.defenseindustrydaily.com>

11 *China aims to double life expectancy* [of sun synchronous satellites from 2-3 years to 4-5 years and geosynchronous satellites from 8 to 15 years] *by 2010*, 20 November 2007, SpaceMart, Space Media Network, n.p., viewed 30 August 2012, <http://www.spacemart.com>.

12 *Space Segment*, 2012, National Coordination Office for Space Based Navigation and Timing, Washington DC, viewed 30 August 2012, <http://www.gps.gov>.

13 *GMV's Director of Space Systems, J Potti, to discuss satellite life extension at Satellite, 2012, Conference in Washington DC,* 2012, GMV, n.p., viewed 30 August 2012, <http://www.gmv.com>.

14 B Springsteen (Team Leader), *Integrated product and process development and case study: development of the F/A-18 E/F*, Institute for Defense Analysis, 1999, Defense Technical Information Center, Fort Belvoir, viewed 28 April 2012, <http://www.dtic.mil>.

15 In June 1774 an experimental submarine, crewed by Mr J Day, submerged at Cattlewater (UK) and did not resurface. (*Maria, Shipwrecks and History in Plymouth Sound*, n.d., The Ships Project, Plymouth, viewed 20 January 2014, <http://www.promare.co.uk/ships/index.htm>.)

16 S Hopper, *Fatalities in human space flight*, (Slide Presentation) n.d., New Mexico State University, Las Cruces, viewed 28 January 2014, <http://astronomy.nmsu.edu >.

17 Between 1957 and 1999 some 4 378 space launches occurred globally with 390 recorded failures, a success rate of 91.9 %. (I-Shih Chang, *Space launch vehicle reliability*, 2012, Aerospace, El Segundo, viewed 22 January 2012, <http://www.aero.com>.)

18 Challenges associated with allocations of radio spectrum for utilisation in outer space have been recognised for decades. For example; T Sandler & W Schulze, 'The economics of outer space,' *Natural Resources Journal*, vol. 21, iss. 2, April 1981, pp. 371 – 393. Located at Space Security, <http://www.spacesecurity.org>

19 *Massive world war mine found in Black Sea*, 25 August 2009, Khaleej Times, United Arab Emirates, viewed 26 April 2012, <http://www.khaleejtimes.com>.

20 *WW II bomb explodes, killing driver*, 3 January 2014, abc News, n.p., viewed 6 January 2014, <http://abcnews.go.com>.

21 *Radioactive fallout from global weapons testing*, updated to 19 August 2010, Centers for Disease Control and Prevention, Atlanta, viewed 24 August 2012, <http://www.cdc.gov>. US Department of Health and Human Services Centres for Disease Control and Prevention and the National Cancer Institute, *Report on the feasibility of a study of the health consequences to the American population from nuclear weapons tests conducted by the United States and other nations*, May 2005, p. 20; Centers for Disease Control and Prevention, Atlanta, viewed 24 August 2012, <http://www.cdc.gov

22 *'Major Findings', Landmine Monitor 2013*, Landmine and Cluster Munition Monitor, Geneva, viewed 28 January 2014, <http://www.the-monitor.org>.

23 *The Problem,* 2012, Cluster Munition Coalition, London, viewed 4 May 2012, <http://www.stopclustermunitions.org>.

24 DG Chandler (foreword by), *The timechart of military history*, 2nd edn, Worth Press &DAG, London, 2000, p. 7.

25 Copyright and related intellectual property resides with the author.

26 Copyright and related intellectual property resides with the author.

Chapter 2

1 B Preston, DJ Johnson, SJA Edwards, *et al. Space weapons Earth wars*, RAND, Santa Monica, 2002, p. 9.

2 Evidence of altitudes attained is:

China's DF15 (CSS-6) missile has been claimed to reach an altitude of 120 km ('*China's growing military power*,' Strategic Studies Institute of the US Army War College, Carlisle, 2002, p. 116.

In a diagram showing the flight profile of an intercontinental ballistic missile with 'booster characteristics similar to those of a US MX missile' an apogee of 1 200 km was included. (A Flax, 'Ballistic Missile Defence,' F Long, DL Hafner & J Boutwell (eds), *Weapons in space*, WW Norton, New York, 1986, p. 55.) Also, ibid. pp. 56-7 discussed the release phase of weapons carried by a multiple independent re-entry vehicle as being from 100 to 1000 km.

It was claimed the late 1980s generation of liquid-fuelled ICBMs reached altitudes of 400 kilometres. R Garwin, 'SDI – a sceptical assessment by an American physicist,' H Brauch & LS

Zuckerman (eds), *Military technology, armaments dynamics and disarmament*, StMartin's Press, New York, 1989, p. 241.) In a subsequent reference, (*loc. cit.*) Garwin claimed 'boosters' were reaching 600 kilometres.

Y Velikhov, R Sagdeev & A Kokoshin (eds). *Weaponry in space*, trans. A Repyev, Mir Publishers, Moscow, 1986, p. 18, published a diagram illustrating the flight profile of an intercontinental ballistic missile with its apogee at approximately 900 km.

B Preston *et al.* in a publication prepared for the United States Air Force, Contract F49642-01-C-0003, stated a ballistic missile's trajectory would normally reach a peak altitude of 'about 1 300 km'.(B Preston, DJ Johnson, SJA Edwards, *et al. Space weapons Earth wars*, op. cit. p. 11, citing R Garwin & H Bethe, 'Anti-ballistic missile systems', *Scientific American*, March 1968, pp. 21 – 30.)

3 D Webb, 'Space weapons: dream, nightmare or reality?', N Borman & M Sheehan (eds), *Securing outer space*, Routledge, New York, 2009, p. 33. (PB Stares recorded 33 such tests by the US alone between October 1959 and November 1984. Stares also cited a publication from 1959 in which (US Army, General) James Gavin claimed it was an urgent necessity to develop a satellite interceptor. (PB Stares, *The militarization of space*, Cornell University Press, Ithaca 1985, p. 261.) Tests were also mentioned by JM Gavin, *War and peace in the space age*, Hutchison, London, 1959, pp. 215 – 16.

4 *Kapton polyimide film*, 2012, DuPont, E I du Pont de Nemours, viewed 17 September 2012, <http://www2.dupont.com>. *Kapton Tape.com*, 2011, Kapton Tape, Torrance, viewed 17 September 2012, <http://www.kaptontape.com>.

5 R Rochus & L Salvador, *Spacecraft thermal control*, November 2011, (PowerPoint presentation), Structural Dynamic Research Group, Department of Aerospace and Mechanical Engineering, University of Liège, Liège, viewed 17 September 2012, <http://www.ltas-vis.ulg.ac.be>.

6 S Aftergood, 'Background on Space Nuclear Power', *Science and Global Security*, vol. 1, iss. 1, 1989, pp. 93 – 107, 105. (Aftergood cited Strategic Defense Initiative Organisation, Reprinted in SDI: *Technology, Survivability and Software*, US Congress, Office of Technology Assessment, OTA-ISC-53, May 1988, p. 142.)

7 United States Congress, Office of Technology Assessment, *Ballistic missile defense technologies*, OTA-ISC-254, US Government Printing Office, Washington DC, September 1985, p. 147.

8 United States Congress, Office of Technology Assessment, *Anti-satellite weapons, countermeasures and arms control*, published as one of two reports jointly titled *Strategic Defenses*, Princeton University Press, Princeton, New Jersey, 1986, p. 73.

9 HN Beveridge, *A review of Project Defender for the Director of Defense Research and Engineering, 1960*, cited by DR Baucom, 'The rise and fall of Brilliant Pebbles', *The journal of Social, Political and Economic Studies*, vol. 29, no. 2, Summer 2004, p. 144, note 2.

10 Raytheon, US manufacturer of the Standard Missile 3, has claimed the kinetic force delivered by that missile at impact as, 'impact is the equivalent of a 10-ton truck [10 160 kg] travelling at 600 mph [965 kph].' (*Standard Missile 3*, n.d, Raytheon, Waltham, viewed 24 March 2013, <http://www.raytheon.com>.) Similar calculations were published in; *RIM-161 SM3* (AEGIS ballistic missile defense), 2013, Global Security, Colorado, viewed 7 January 2014, <http://www.globalsecurity.org>. Details of the design of the SM-3 warhead were published by (US) Naval Surface Warfare Center. (D Houchins, *Hazard assessment testing of the SM-3 Block 1A missile*, 13 March 2007, Defense Technical Information Center, Fort Belvoir, viewed 7 January 2014, <http://www.dtic.mil>.)

11 The first photovoltaic cells, invented by Chapin, Fuller &, Pearson in 1954 at the Bell Laboratory (US), had an energy conversion efficiency of 4 %. In 1960 Hoffman electronics achieved 14 % and by 1994 the (US) National Renewable Energy Laboratory invented the gallium arsenide cell which could convert solar energy with an efficiency of 30 %. (*The story of solar*, c. 2001. US Department of Energy, Washington DC, viewed 15 December 2011, <http://www1.eere.energy.gov>.)

12 *Power systems, radioisotope thermoelectric generator*, 4 September 2013, National Aeronautics and Space Administration, Washington DC, viewed 12 October 2014, <http://solarsystem.nasa.gov>.

13 R Bulkeley, 'Nuclear power in space – technology beyond control', HG Brauch (ed.), *Military technology, armaments dynamics and disarmament,* St Martins Press, New York, 1989, pp. 183 – 4.

14 *Nuclear reactors for space*, November 2011, World Nuclear Association, London, viewed 29 March 2012, <http://www.world-nuclear.org>. The same reactor was acknowledged by; RX Lenard, 'Review of reactor configurations for space nuclear electric propulsion and surface power considerations', in C Bruno (ed.), *Nuclear space power and propulsion systems*, American Institute of Astronautics and Aeronautics, Reston, 2008, p. 178.

15 R Bulkeley, Nuclear power in space, *op. cit.* pp. 184 – 5 and 190. Some details of the nuclear reactor are in; *Protocol Between the Government of Canada and the Government of the Union of Soviet Socialist Republics,* done on April 2, 1981 with statement of the Canadian claim [for damage caused by the re-entry of satellite *Cosmos 954*].

16 Compound annual growth rate of revenue to the PRC nuclear energy sector between 2007 and 2011 was 8.8 % per annum. A growth of 3.5 % per annum for the period 2011 to 2016 was forecast. (*Research and Markets: Nuclear Energy in China*, 7 August 2012, Business Wire, New York Office, viewed 22 January 2013, <http://www.businesswire.com>.)

17 '*The role of nuclear power and nuclear propulsion in the peaceful exploration of space,* 2005, p. 5. Located at International Atomic Energy Agency, Vienna, <http://www.iaea.org>.

18 RX Lenard, 'Review of reactor configurations for space nuclear electric propulsion and surface power considerations', C Bruno (ed.), *Nuclear space power and propulsion systems, op. cit.* p. 178.

19 For example, the IRAS satellite of 1983 could detect thermal signatures as cool as -258° C. (*IRAS: Mapping the infrared sky*, 2011, National Aeronautics and Space Administration, Washington DC, viewed 19 August 2011, <http://www.nasa.gov.>.)

20 More accurately, the probable value is: *1360.8 +/- 0.5 Wm²*. G Kopp & JL Lean, 'A new, lower value of total solar irradiance: evidence and climate significance', *Geophysical Research Letters*, vol. 38, iss. 1, 2011, L01706.

21 The claimed efficiency of any photovoltaic cell in converting sunlight into electrical power should be read as indicative rather than specific, except when a significant number of factors affecting efficiency have been both described and measured and the attributes of the 'sunlight' employed are disclosed. In 2008, a solar cell has achieved a conversion efficiency of 40.8 %. (J Geisz, DJ Friedman, JS Ward, *et al.* '40.8% efficient triple junction solar cell with two independently metamorphic functions', 23 September 2008, *Applied Physics Letters*, vol. 93, iss. 12.) A subsequent report claimed *a three-layered cell* had converted 43.5 % of the energy of sunlight into electrical power. (*Award winning PV cell pushes efficiency higher*, 28 December 2012, National Renewable Energy Laboratory, US Department of Energy, Washington DC, 9 February 2014, <http://www.nrel.gov>.)

22 B Cook, *An introduction to fuel cells and hydrogen technology*, 2001, FuelCellStore, 2012, n.p., viewed 23 January 2013, <http://www.fuelcellstore.com>. A range of relevant information and data is available at Fuel Cells Technology Office, a component of US Department of Energy, 2012, Washington DC, viewed 25 January 2012, <http://www1.eere.energy.gov>. *Fuel cells; a better energy resource,* 2010, Glenn Research Center, National Aeronautics and Space Administration, Cleveland, viewed 25 January 2012, <http://www.nasa.gov>. *Fuel cell efficiency*, World Energy Council, 2012, London, viewed 25 January 2012, <http://www.worldenergy.org>.

23 W Ley & R Röder, 'Electrical Power Supply', W Ley, K Wittmann & W Hallmann (eds), *Handbook of space technology*, John Wiley, Chichester, 2009, p. 244, Table 4.2.1 Technical data of the Space Shuttle orbiter fuel cells.

24 Efficiency factors vary between fuel-cell types, and are affected by a number of effects including temperature, gas concentration, pressure and voltage. Accurate efficiency determinations therefore tend to be time, place and case specific. (*Efficiency and open circuit voltage*, 2012; National Kaohsiung University of Applied Sciences, Kaohsiung, viewed 25 January 2012, <http://eng.kuas.edu.tw>.)

Sources cite different values for the energy-density-by-mass (low-heat-value index) of hydrogen. A value of 119.95 MJ/kg is here accepted, it being the mean of five observations recorded by E Tzimas, C Filiou, S Peteves & J Veyret, *Hydrogen storage*, Joint Research Centre, European Commission, Petten, 2003, p. 16. Located at Institute for Energy, Joint Research Centre of the European Commission, Brussels, viewed 26 January 2012, <http://ie.jrc.ec.europa.eu> (website later amended to <http://iet.jrc.ec.europa.eu/>.) 119.95 MJ/kg converts to 33.319 kWh (conversion factor of 3.619553738). Author.

25 *The Physics Hypertextbook*, 2011, G Elert, n.p., viewed 27 January 2012, <http://www.physics.info>.

26 The energy density mass shown for hydrogen (141 MJ/kg) varies from the value of 119.95 used in the calculations of fuel requirements in the preceding text paragraph. The value 141 represents hydrogen at High Heat Value, as indicated in the table, whereas the 119.95 used for the fuel requirement calculation in the text represents hydrogen at Low Heat Value (LHV).

27 W Ley & R Röder, 'Electrical Power Supply,' *op. cit.* p. 244, Table 4.2.1 Technical data of the Space Shuttle orbiter fuel cells.

28 B Cook, *An introduction to fuel cells and hydrogen technology*, *op. cit.* s. 3.2.2, figure. 7.

29 WWS Wong & J Fergusson, *Military space power A guide to the issues*, Praeger, Santa Barbara, 2010, p. 74. T Malik, *Prototype satellites demonstrate in-orbit refuelling*, 4 April 2007, TechMediaNetwork, New York, viewed 26 January 2012, <http://www.space.com>.

30 Y Chen, Y Huang & X Chen, 'Development of simulation testbed for autonomous on-orbit servicing technology,' *Robotics, Automation and Mechatronics* (2011 IEEE Conference), September 2011) pp. 148 – 153.

31 US Human Spaceflight Plans Committee, *Seeking a human spaceflight program worthy of a great nation*, 2009, Washington DC, (the Augustine Report) p. 100 – 102. J Young, *A value proposition for lunar architectures utilising on-orbit propellant refuelling*, 2009, Georgia Institute of Technology, Atlanta, viewed 24 January 2012, <http://www.ssdl.galtech.edu>.

Chapter 3

1. '*The greater kill radius of a nuclear warhead might compensate for inability to achieve a close approach*,' Office of Technology Assessment, *Ballistic missile defense technologies*, *op. cit.* p. 158
2. *Chart of Strategic Nuclear Bombs*, c. 2012, Strategic-Air-Command, Omaha, viewed 21 September 2012, <http://www.strategic-air-command.com>.
3. D Wright, L Grego & L Gronlund, The p*hysics of space security A reference manual,* American Academy of Arts and Sciences, Cambridge, 2005, p. 77.
4. JC Moltz, *The politics of space security: strategic restraint and the pursuit of national interests,* Stanford University Press, Stanford, 2008, p. 97, citing N Christofilos, the information from Christofilos having been obtained from an interview at the Stanford Linear Accelerator on 9 March 2006.
5. H Hoerlin, *United States high-altitude test experiences*, 1976, Los Alamos Scientific Laboratory, p. 25; located at Federation of American Scientists, <http://www.fas.org>.
6. PB Stares, *Space weapons and US strategy*, Croom Helm, London, 1985, pp. 136 – 140.
7. United States Congress, Office of Technology Assessment, *Ballistic missile defense technologies*, *op. cit.* p. 70, diagram, fig. 4.1, U.S.S.R. ICBMs.

8. *Russian sub fires ballistic missile in trial,* 28 July 2011, Global Security Newswire, n.p., viewed 30 July 2011, <http://gsn.nti.org>.

9. *New ICBM planned in Russia,* 19 July 2011, Global Security Newswire, n.p., viewed 30 July 2011, <http://gsn.nti.org>.

10. R Pandit, *With China in mind, India tests new-generation Agni missile with high kill efficiency,* 16 November 2011, The Times of India, n.p., viewed 16 March 2012, <http://timesofindia.indiatimes.com>.

11. *India Profile,* Missile Overview, October 2009, Nuclear Threat Initiative, Washington DC, viewed 14 December 2011, <http://www.nti.org>.

12. Kapila recorded the *Shaheen II* missile as having a maximum range of 2 500 km and capable of transporting a 1 000 kg payload. (S Kapila, *Pakistan's ballistic missile arsenal* (Paper 48), 28 September 2000, South Asian analysis, Noida,<http://www.southasiananalysis.org>.) Other sources assess the apparently equivalent *M-18* missile as having a 1 000 km range with a 500 – 800 kg payload. (For example, *Weapons of Mass Destruction, Shaheen II/Hatf-6/Ghaznaf,* n.d., Federation of American Scientists, Washington DC, viewed 29 March 2012,<http://www.fas.org>.)

13. J Arya, 'Notes on the Nodong-Ghauri missile handover ceremony at Khan Research Labs Complex', *Bharat Rakshak Monitor*, vol. 5, iss. 4, January- February 2003. Located at, Bharat Rakshak, (Official Indian Armed Forces Site), <http://www.bharat-rakshak.com> .

14. *M-4/M-45*, n.d., Federation of American Scientists, Washington DC, viewed 29 March 2012, <http://www.fas.org>.

15. A Federation of American Scientists document recorded nine Chinese nuclear-capable missiles, all of which appeared to have sufficient range to enable the launch of a nuclear weapon into outer space. The missiles were: (US/NATO designations) DF-3A, DF-4, DF-5A, DF-21-A, DF-31, DF-31A, JL-1, JL-2 and Hong-6. (*Nuclear Weapons,* Table: Estimated Chinese Nuclear Forces, 2006, Federation of American Scientists, Washington DC , viewed 29 March 2012,<http://www.fas.org>.)

16. R Norton-Taylor, *Britain's nuclear arsenal is 225 warheads, reveals William Hague*, 26 March 2010, Guardian News and Media, London, viewed 11 January 2012, <http://www.guardian.co.uk>. (The article claimed 160 such weapons were 'operationally ready' but no further details were provided. Additional information was not found.)

17. WD Farr, *The third temple's holy of holies: Israel's nuclear weapons*, United States Air Force Counterproliferation Center, Maxwell Air Force Base, 1999. Located at Federation of American Scientists, Washington DC, <http://www.fas.org>.

18. J Heller, *Israel test fires ballistic missile: Israel radio*, 2 November 2011, Jewish Journal of Greater Los Angeles, Los Angeles, n.p., viewed 14 December 2011, <http://www.jewishjournal.com>.

19. *Musadan*, 2011, Missile Threat, Claremont Institute, Claremont, viewed 14 December 2011, <http://www.misssilethreat.com>.

20. *Atomic Energy Basic Act* (No. 186 of 1955) effective 19 December 1955, Article 2, incorporating revisions to 3 December 2004, (translated from Japanese to English 31 March 2010, translator not identified). Located at Nuclear Safety Commission, Tokyo, viewed 12 January 2012, <http://www.nsc.go.jp>; also; *Japan: Nuclear weapons program,* 2012, Global Security, Colorado, viewed 12 January 2012, <http://www.globalsecurity.org>.

21. M Fackler, *Chinese patrol ships pressuring Japan over islands*, 2 November 2012, The New York Times, New York, viewed 27 November 2012, <http://www.nytimes.com>. *Abe says Japan's pacifist constitution may be revised by 2020*, 2 January 2014, Space War, Space Media Network, n.p., viewed 3 January 2014, <http://www.spacewar.com>.

22. Y Zaitsev, *Russia begins elbowing Ukraine out from Brazil's space program*, 18 September 2008, RIA Novosti. Located in Space Daily, n.p., viewed 3 January 2014, <http://www.spacedaily.com>. D Messier, *Brazil scales back launch vehicle plans,* 10 February 2013, Agência Espacial Brasileira via Gazeta do Povo, n.p., <http://www.parabolicarc.com>. A Svitak, *Ukraine, Brazil prepare for 2015 Cyclone 4 launch*, n.d., Aviation Week & Space Technology New York, viewed 3 January 2014, <http://www.aviationweek.com>.

23. B Preston, DJ Johnson, SJA Edwards, *et al. Space weapons Earth wars*, *op. cit.* p. 12.

24. *Review of scientific and astronautical research and development in the Department of Defense,* Staff Report, 87th Congress, 1st Session, 1961, p. 54.

25. United States Congress, Office of Technology Assessment, *Anti-satellite weapons, countermeasures and arms control*, *op. cit.* p. 55, p. 58.

26. D Webb, 'Space weapons: dream, nightmare or reality?' *op. cit.* p. 29.

27. H Caldicott & C Eisendrath, *War in heaven*, The New Press, New York, 2007, p. 73, citing J Singer, *Space News*, 18 April 2005.

28. H Caldicott & C Eisendrath, *War in heaven, op. cit.* pp. 77 - 78, citing T Weiner, *Air Force Seeks Bush's Approval for Space Weapons Program*, New York Times, 18 May 2005. M Hoey (Research Associate, Institute for Defense Disarmament Studies), *Military space systems: the road ahead*, 22 October 2005, p. 9. Located at The Space Review <http://www.thespacereview.com>.

29. *Exoatmospheric kill vehicle plays key role in latest missile defense test*, 20 August 2010, Space War, Space Media Network, n.p., viewed 20 August 2010, <http://www.spacewar.com>.

30. J Lewis, 'Hit to kill' and the threat to space assets', *Celebrating the space age* (Conference Report, 2007), United Nations Institute for Disarmament Research, Geneva, 2007, p. 149.

31. A Krishnan, *Indian official says anti-satellite weapon within reach*, 7 March 2011, Aviation Week & Space Technology, New York, viewed 5 February 2012, <http://www.aviationweek.com>. B Gopalaswamy & G Kampani, *Piggybacking anti-satellite technologies on ballistic missile defense: India's hedge and demonstrate approach*, 19 April 2011, Carnegie Endowment, Washington DC, viewed 5 February 2012, <http://www.carnegieendowment.org>.

32. *Hermes 450*, June 2003, Israeli Weapons.com Ltd, n.p., viewed 15 June 2014, <http://www.israeli-weapons.com/weapons/aircraft/uav/hermes_450.html>.

33. For deployment of SM-3 missile, Fact sheet on UN missile defense policy. *A 'phased adaptive approach' for missile defense in Europe,* 17 September 2009, White House Office of the Press Secretary, Washington DC, viewed 24 March 2013, <http://www.whitehouse.gov>.

34. Space-mine concepts were referred to by: B Preston, DJ Johnson, *et al. Space weapons Earth wars, op. cit.* p. 9. WWS Wong & J Fergusson, *Military space power,* op. cit. pp. 8 – 9. B Chapman, *Space warfare and defense*, ABC-CLIO, Santa Barbara, 2008, p. 171. Estimate of orbiting payload numbers from: *UCS satellite data base,* 31 January 2014, Nuclear Weapons and Global Security, Union of Concerned Scientists, Cambridge, viewed 28 May 2014, <http://www.ucsusa.org/nuclear_weapons_and_global_security/solutions/space-weapons/ucs-satellite-database.html

Chapter 4

1. United States Congress, Office of Technology Assessment, *Anti-satellite weapons, countermeasures and arms control, op. cit.* p. 69. Subsequently D Beason listed 60 categories of laser. (D Beason, *The e-bomb,* Da Capo Press, Cambridge, 2006, Appendix, Types of Lasers, pp. 217 – 219.)

2. United States Congress, Office of Technology Assessment, *Ballistic missile defense technologies,* published as one of two reports jointly titled *Strategic Defenses,* Princeton University Press, Princeton, New Jersey, 1986, p. 149.

3. ibid. p. 150.

4. J Hecht, *Military mega-lasers are too hot to handle,* 10 July 2009, New Scientist, iss. 2716, viewed 16 June 2014, <http://www.newscientist.com/article/mg20327166.000-military-megalasers-are-too-hot-to-handle.html>.

5. United States Congress, Office of Technology Assessment, *Anti-satellite weapons, countermeasures and arms control, op. cit.* p. 53.

6. United States Congress, Office of Technology Assessment, *SDI Technology, Survivability and Software,* Princeton University Press, Princeton, 1988, pp. 130 -132.

7. P Boffrey, W Broad, *et al. Claiming the heavens,* Times Books, New York, 1988, p. 82.

8. A September 1985 demonstration at White Sands was described by an observer in the following terms:

   [The laser] ... *' looks more like a giant diesel engine than a futuristic weapon and requires dozens of delicate mirrors, 370 people and 9 000 gallons of water to operate, was then focussed for several seconds on the Titan* [missile] *until the cables holding it up snapped and gave the impression that the missile was exploding ... In congressional testimony a month later General Abrahamson maintained that the White Sands test 'demonstrated graphically the lethality of this technology'.*

J Tirman, 'The politics of star wars', J Tirman (ed.), *Empty promise: the growing case against star wars*, Union of Concerned Scientists, Boston, 1986, p. 17.

9. V Dvorkin, 'Space weapons programs', A Arbatov & V Dvorkin (eds), *Outer space weapons, diplomacy and security,* Carnegie Endowment for International Peace, Washington DC, 2010,. p. 36.

10. V Dvorkin, 'Space weapons programs', *op. cit.* p. 36, citing assessments made by a team of experts. (No data or methodology was provided and the experts' qualifications were not disclosed.) Readers were referred to Molchanov, *Nuclear Proliferation: New Technologies* pp. 196 – 228. This test was mentioned also by W WS Wong & J Fergusson, *Military space power, op. cit.* p. 5, citing F Sietzen, '*Laser hits orbiting satellite in beam test*', Space Daily, international, 20 October 1997. A similar account is contained in D Webb, 'Space weapons: dream, nightmare or reality?' *op. cit.* p. 29.

11. Office of the Under Secretary for Defense for Acquisition Technology and Logistics, *Defense Science Task Force on High Energy Laser Weapon Systems Application*, US Department of Defense, Washington DC, June 2001, p. xi, p. 19; located at scribd, <http://www.scribd.com>.

12. I Kramnik, *How real is the threat of laser weapons?,* 24 February 2010, Space War, Space Media Network, n.p., viewed 1 February 2012, <http://www.spacewar.com>. Also, D Webb, 'Space weapons: dream, nightmare or reality?' *op. cit.* p. 29.

13. *Airborne laser test bed successful in lethal intercept experiment,* 11 February 2010, Missile Defense Agency, Arlington, viewed 19 February 2014, <http://www.mda.mil/news/10news0002.html>.

14. *YAL-1A airborne laser test bed,* 2012, Global Security, Colorado, viewed 7 March 2012, <http://www.globalsecurity.org>. The maximum payload of a C 17 aircraft is approximately 77 tonnes. (*C-17 Globemaster III features,* 2012, Global Aircraft, n.p., viewed 7 April 2012, <http://www.globalaircraft.org>.)

15. References are:

    Net chemical efficiency of 23.4 %. (M Endo, S Nagatomo, S Takeda, *et al.* 'High efficiency operation of a chemical oxygen-iodine laser using nitrogen as a buffer gas', IEEE *Journal of Quantum Electronics*, vol. 34, iss. 3 (March 1998).)

    Chemical efficiency 32.4 %. (M Endo, T Osaka & S Takeda, 'High efficiency operation of chemical oxygen-iodine laser using a streamwise vortex generator', *Applied Physics Letters*, vol. 84, iss. 16, 2004.)

    Chemical efficiency of 25 %. (F Wani, M Endo & T Fujioka, 'High-pressure, high-efficiency operation of a chemical oxygen-iodine laser', *Applied Physics Letters*, vol. 75, iss. 20, (1999).)

    Maximum overall efficiency of 40 %. (S Yoshida, M Endo, T Sawano, *et al.* 'Chemical oxygen iodine laser of extremely high efficiency', *Journal of Applied Physics*, vol. 65, iss. 870, (1989).)

16. *X-ray laser pulses of unprecedented energy and brilliance produced at SLAC*, 21 April 2009, National Accelerator Laboratory, Menlo Park, viewed 13 October 2011, <http://www.slac.stanford.edu>.

17. *Fusion: X-ray laser zaps solid to 2 million degrees*, 25 January 2012, Space Daily, Space Media Network, n.p., viewed 28 January 2012, <http://www.spacedaily.com>.

18. WT Silfvast, 'Lasers', M Bass (ed.), *Handbook of Optics*, vol. 1, McGraw-Hill, New York, 1995, p. 11.33

19. A Mandi, 'XeF (B-X) long-pulse-length laser studies', *Journal of Applied Physics*, vol. 71, iss. 4, 1992. H von Bergmann, U Rebhan & U Stamm, 'Design and Technology of Excimer Lasers', D Basting and G Marowsky (eds), *Excimer Laser Technology*, Springer, Berlin, 2005, p. 47. WT Silfvast, 'Lasers', in; M Bass (ed.), *Handbook of Optics, op. cit.* p. 11.33 (claimed between 1 and 5 % efficiency).

20. RJ Jensen, 'Los Alamos Krypton Fluoride Laser Program', *Laser and Particle Beams*, vol. 4, iss. 01, 1986, pp. 3 -16.

21. B Anderberg & M Woldbarsht, *Laser weapons the dawn of a new military age*, Plenum Press, New York, 1992. pp. 107 -137. Located at Laser Stars Org, <http://www.laserstars.org>. Also, *Ground based laser*, 2012, Global Security, Colorado, viewed 1 February 2012, <http://www.globalsecurity.org> .

22. D Mudge, M Ostermeyer, DJ Ottaway, *et al.* 'High-power Nd:YAG lasers using stable-unstable resonators', *Classical and Quantum Gravity*, vol. 19, no. 7, 2002, pp. 1783 - 1792.

23. SJ Mc Naught, CP Asman, H Injeyan, *et al. 100 kW coherently combined Nd:YAG MOPA* [master oscillator power amplifier] *array*, conference paper, Frontiers in Optics, San Jose October 2009. Located at The Optical Society of America, Washington DC, <http://www.opticsinfobase.org>.

24. *Scientific assessment of high-power free-electron laser technology*, 2008, p. 11, Committee on a Scientific Assessment of High-Power Free-Electron Laser Technology for Naval Applications, National Academies Press, National Academy of Sciences, Washington DC, <http://www.nap.edu>.

25. *Scientific assessment of high-power free-electron laser technology, op. cit.*

26. CA Allen, *Integrating the FEL into an all electric ship*, Naval Post Graduate School, Monterey, 2007, p. 20.

27. *Scientific assessment of high-power free-electron laser technology, op. cit.*

28. *Free electron laser*, n.d., United States Office of Naval Research, United States Navy, Arlington, viewed 13 October 2011, <http://www.onr.navy.mil>.

29. S Ackerman, *Power down: Senate zaps Navy's superlaser, railgun,* 17 June 2011, Wired.com, San Francisco, viewed 1 February 2012, <http://www.wired.com>.

30. *Military Analysis Network, Palletized Load System (PLS)*, 3 January 2000, Federation of American Scientists, Washington DC, viewed 22 December 2010, < http://www.fas.org >.

31. C Kopp, *High energy laser directed energy weapons*, Technical Report APA-TR-2008-0501, 2008, Air Power Australia, n.p., <http://www.ausairpower.com>.

32. C Kopp, *High energy laser directed energy weapons, op. cit.*

33. I Kramnick, *How real is the threat of laser weapons?*, 24 February 2010, Space War, Space Media Network, n.p., viewed 1 February 2012, <http://www.spacewar.com>.

34. M Peach, *Conference expects slow transition to laser weapons*, 7 March 2012, Optics.org, Cardiff, viewed 29 January 2012, <http://www.optics.org>.

35. V Raghuvanshi, *Defense News*, 25 August 2010, Defense News, n.p., viewed 19 December 2011, <http://www.defensenews.com>. P de Selding, *India developing anti-satellite spacecraft*, 11 January 2010, TechMediaNetwork, New York, viewed 19 December 2011, <http://www.space.com>.

36. G Leopold, *Israel test laser weapon*, 7 June 1999, Electronic Ecosystem, n.p., viewed 13 September 2012, <http://www.eetimes.com>.

37. *Tactical high energy laser*, 2012, Missile Threat, Claremont, viewed 13 September 2012, <http://www.missilethreat.com>.

38. *Iran builds new weapons with wave of the wand*, 11 November 2010, Pravda Ru, Moscow, viewed 13 September 2012, < http:///www.english.pravda.ru >.

39. J Stupl, N Faber, C Foster, *et al. LightForce photon-pressure collision avoidance: efficiency assessment based on an entire catalog of space debris*, NASA Ames Research Center, Moffett Field, 2012.

40. United States Congress, Office of Technology Assessment, *Anti-satellite weapons, countermeasures and arms control, op. cit.* p. 71. H Bethe, J Boutwell & RL Garvin, 'BMD technologies and concepts in the 1980s', in F A Long, D Hafner & J Boutwell (eds), *Weapons in Space*, Norton, New York, 1986, pp. 61 – 62.

41. United States Congress, Office of Technology Assessment, *Anti-satellite weapons, countermeasures and arms control, op. cit.* p. 71, 73.

42. PM Boffrey, WJ Broad, *et al. Claiming the heavens, op. cit.* p. 92.

43. The description of the beam as 'neutral' is a generalisation. A description of the experiment included the following information: 'The neutralizer had been programed to produce a slightly net negative current of approximately 1 mA with individual species currents of ......Late in the flight excess xenon was injected into the beam producing a net positive beam current of + 1.4 mA.' (PG O'Shea, TA Butler, MT Lynch, *et al. A linear acceleration in space – the beam experiment aboard a rocket*, Los Alamos National Laboratory, September 1990. Located at Federation of American Scientists, <http://www.fas.org>.)

44. *Beam experiment aboard a rocket (BEAR)*, n.d., Global Security, Colorado, viewed 9 February 2012, <http://www.globalsecurity.org>.

45. GJ Nunz, *BEAR (Beam Experiments Aboard a Rocket) project final report*, vol. 1, Project summary, Bear Project Staff, Strategic Defense Initiative Organization Directed Energy Office, Washington DC, 1990, Table IV, BEAR vehicle geometry and mass properties.

46. PG O'Shea, TA Butler, MT Lynch, *et al. A linear acceleration in space – the beam experiment aboard a rocket', op. cit.*

47. United States Congress, Office of Technology Assessment, *Anti-satellite weapons, countermeasures and arms control, op. cit.* p. 73.

48. ibid.

49. RL Garwin, *Weapons in space; Are we on the verge of a new arms race?* A talk to the AAAS Symposium on Trends in Strategic Weapons and Doctrines and their Implications for Arms Control, Toronto. 4 January 1981. Located at Federation of American Scientists, <http://www.fas.org>.

50. United States Congress, Office of Technology Assessment, *Anti-satellite weapons, countermeasures and arms control, op. cit.* p. 54.

51. US Department of Energy (Report), A*ccelerators for America*, 2010, US Department of Energy, Washington DC, p. 60. For more detail; A Todd, M Nightingale, T Yule, *et al. The continuous wave deuterium demonstrator (CWDD) design and status*, 1993, Advanced Energy Systems, Medford, viewed 9 February 2012, <http://www.aesys.net> The Argonne Accelerator History Document Collection, *CWDD, Continuous wave deuterium demonstrator*, n.d., Argonne Accelerator Institute, Chicago, viewed 9 February 2012, <http://www.aai.anl.gov>.

52. *Accelerators for America, op. cit.* p. 67.

53. ibid. p. 61.

54. ibid. p. 64.

55. Annual report to Congress, *Military and security developments involving the People's Republic of China*, US Department of Defence, Washington DC, 2011, p. 37. Located at; US Department of Defense, <http://www.defense.gov>.)

56. *Shanghai particle accelerator to open its doors for business*, 30 April 2009, Xinhua News Agency, Beijing, viewed 8 July 2014, <http://news.xinhuanet.com/english/2009-04/30/content_11287049.htm>.

57. D Overbye, *Price of next big thing in physics: 6.7 billion,* 8 February 2007, The New York Times, New York, viewed 10 October 2011, <http://www.nytimes.com>.

58. *Kali (Kilo Ampere Linear Injector), Indian Beam Weapon?* n.d., Pakistani Defence Forum, Indianapolis, viewed 4 February 2012, <http:///www.forum.pakistanidefence.com>.

59. *Indian secreat* [sic] *projects: do they really exist?* 2012, Indian Defence, n.p., viewed 4 February 2012, <http://www.indiandefence.com>.

60. *Ballistic Missile Defense Review Repor*t, February 2010, US Department of Defense, Washington DC, viewed 15 March 2012, <http://www.defense.gov>.

Chapter 5

1. L Hui, *Development of foreign high-powered microwave weapons and prospects of future applications in space-based target defense and air defense*', National Air Intelligence Center (NAIC-ID (RS) T-0617-95) 8 March 1996, Defense Technology Information Center, Fort Belvoir, viewed 6 February 2012, <http://www.dtic.mil>.
2. WWS Wong & J Fergusson, *Military space power, op. cit.* p. 85.
3. V Dvorkin, 'Space weapons programs,' in A Arbatov & V Dvorkin (eds), *Outer space weapons, diplomacy and security, op. cit.* p. 39.
4. V Dvorkin, 'Space weapons programs,' *op. cit.* p. 38.
5. *BBC World Service short wave radio blocked in China*, 25 February 2013, BBC, London, viewed 10 January 2014, <http://www.bbc.com>.
6. *China's jam-proof satellite going up next year*, 4 March 2004, Pakistan Defence, Indianapolis, viewed 10 January 2014, <http://forum.pakistanidefence.com/index.php?showtopic=27311>.
7. R Briel, *Eutelsat condemns Iranian jamming*, 4 October 2012, Broadband TV News, Cambridge, viewed 10 January 2014, http://www.broadbandtvnews.com/2012/10/04/eutelsat-condemns-iranian-jamming
8. For example, J Quain, *GPS jammers can wreak havoc, cover crimes,* 22 March 2011, The Associated Press, New York, viewed 6 February 2012, <http://www.msnbc.msn.com>.
9. N Lewer, *Bradford non-lethal weapons research project,* June 1988, University of Bradford, Bradford, viewed 14 September 2012, <http://www.brad.ac.uk>.
10. *Microwave weapon designed to render electronics useless*, March 2011, National Defense Magazine, 2012, National Defense Industry Association, Arlington, viewed 14 September 2012, <http://www.nationaldefensemagazine.org>.
11. D Wright, L Grego & L Gronlund, *The physics of space security, op. cit.* pp. 132 -133.
12. *Frying tonight,* 15 October 2011, The Economist, n.p., viewed 14 September 2012, <http://www.economist.com>
13. C Kopp, *Russian/Soviet point defence weapons,* updated April 2012, Air Power Australia, n.p., 2012, viewed 14 September 2012, <http://www.ausairpower.net>.
14. *DRDO works on high power electromagnetic weapon and electronics hardening techniques,* 30 July 2011, Frontier India News Network, Maharashtra, viewed 14 September 2012, <http://frontierindia.net>.
15. S Weinberger, 'Microwave weapons; wasted energy,' *Nature,* no. 489, iss. 7415, September 2012,. pp. 198 -200. Located at; Nature, <http://www.nature.com>.

16. *Transiting from air to space. The North American X-15, Section VI, Research at the edge of space*, n.d., National Aeronautics and Space Administration, Washington DC, viewed 24 February 2011,<http://www.nasa.gov>.

17. *Space shuttle era facts*, n.d., National Aeronautics and Space Administration, JFK Space Center Florida, viewed 6 April 2014, <http://www.nasa.gov>.

18. S Pace, *National Aerospace Plane Project; principal assumptions, findings and policy options*, RAND, Santa Monica, 1986.

19. H Caldicott & C Eisendrath, *War in heaven, op. cit.* p. 82, citing W Pincus, *'Pentagon has Far-Reaching Defense Spacecraft in Works'*, Washington Post, 16 March 2005. M Hoey, *Military Space Systems: The Road Ahead, op. cit.* pp. 14 – 15.

20. *Air Force officials launch Atlas V carrying X-37B orbit test vehicle*, 23 April 2010, United States Air Force, n.p., (US), viewed 13 May 2012, <http://www.af.mil>. On 11 December 2013 the X-37B had spent one year on orbit during one of its sorties into outer space. (M Jones, *X-37B marks one year on orbit*, 4 December 2013, Space War, Space Media Network, n.p., viewed 28 December 2013, ,http://www.spacewar.com>.)

21. R Fisher, *China's space plane program*, 27 July 2011, International Assessment and Strategy Center, Alexandria, viewed 20 January 2012, <http://www.strategycenter.net>. These claims received hesitant support from one Chinese news source. (*China's space plane succeeds in test fly*, 29 April 2010, Hao Hao Report, n.p., viewed 19 January 2012, <http://www.haphaoreport.com >.)

22. S Mohanty, *Indian space program*, June 2008, Earth 2 Orbit, Antrix Corporation (Commercial arm of Indian Department of Space), Bangalore, viewed 20 January 2012, <http://www.earth2orbit.com>.

23. S Clark, *History-making Japanese space mission ends in flames*, 1 November 2009, Spaceflight Now, n.p., viewed 20 January 2012, <http://www.spaceflightnow.com>.

24. D Gayle, *Japanese scientists fear spacecraft blueprint stolen after networks penetrated by virus*, 18 January 2012, Mail Online, London, viewed 20 January 2012, <http://www.dailymail.co.uk>

25. S Canright, *Astronaut requirements*, 2010, National Aeronautics and Space Administration, Washington DC, viewed 15 June 2011,< http://www.nasa.gov/audience/students>. In January 2012 NASA reported that a Russian *Progress* resupply vehicle (an autonomous version), had visited the International Space Station and, following departure had deployed a satellite. (P Harding, *Russian Progress M-14M docks – M-13M de-orbits following satellite deploy*, 27 January 2012, NASA Spaceflight, n.p., viewed 21 January 2014, <http://www.nasaspaceflight.com/2012>.)

26. JR Minkel, *Is the International Space Station worth USD 100 billion?* 1 November 2010, Space, TechMediaNetwork, New York, viewed 28 November 2012, <http://www.space.com>.

27. *Russia to build world's first commercial scramjet*, 20 January 2008, Pravda. Ru, Moscow, viewed 15 December 2011, <http://engforum.pravda.ru>; *Commissioning of scramjet combustion flight experiment*, 20 March 2006, Japan Aerospace Exploration Agency, Chofu-city, viewed 15

December 2011, <http://www.jaxa.jp>.

28. C Covault, *China developing scramjet propulsion*, 2 September 2007, Aviation Week & Space Technology, New York, viewed 15 December 2011, <http://www.aviationweek.com>. *India joins elite nations with scramjet technology*, 11 January 2006, Silicon India, Murugeshpalya, viewed 15 December 2011, <http://www.siliconindia.com>. RC Harding, *Space policy in developing countries*, Routledge, London, 2013, p. 107.

29. *China tested hypersonic missile vehicle: US official*, 16 January 2014, Space Travel, n.p., viewed 18 January 2014, <http://www.space-travel.com>.

30. DD Day & RG Kennedy, *Soviet star wars*, January 2010, Air and Space, Smithsonian, Washington DC, viewed 20 December 2011, <http://www.airspacemag.com>.

31. MT Eldridge, KH McKechnie & RM Hefley (Inventors), *Satellite signature suppression shield*, filed 14 March 1990, US Patent 5,345,238, 6 September 1994, available at, <http://www.patents.com>.

32. Y Velikhov, R Sagdeev and A Kokoshin (eds), Weaponry in space, *op. cit.* p. 102. *A chameleon in the physics lab*, 29 October 2013, Space Daily, Space Media Network, n.p., viewed 27 December 2012, <http://www.spacedaily.com>. *Thin, active invisibility cloak demonstrated for the first time*, 12 November 2013, Phys Org, n.p., viewed 29 December 2013, <http://phys.org/news/2013-11-thin-invisibility-cloak.html>.

33. P Alitalo, H Kettunen & S Tretyakov, 'Cloaking a metal object from an electromagnetic pulse: a comparison between various cloaking techniques', *AIP Journal of Applied Physics*, vol. 107, iss. 3. (February 2010). (Ideal cloaking appeared possible only at a single frequency. Other cloaking limitations were identified. Author.)

34. 'On-orbit deception', 7 June 2011, *Air Force Magazine*. Located at Air Force Association, Arlington, <http://www.airforcemag.com> .

35. IPS Radio and Space Services, *Electrons damage satellites*, 2011, Ionospheric Prediction Service, Radio and Space Services, Space Weather Branch, Bureau of Meteorology, Melbourne, viewed 14 October 2011, <http://www.ips.gov.au>.

36. Data from lecture by Dr O Giersch to 12th Australian Space Science Conference, held Melbourne, September 2012.

37. JA Kennewell & B-N Vo, *An overview of space situational awareness*, 2013, authors' publication, <http://ba-ngu.vo-au.com/vo/KV_SSA_FUSION13.pdf>.

38. For one example, *Small is beautiful*, 30 June 2011, Defense Industry Daily, n.p., viewed 7 March 2013, <http://www.defenseindustrydaily.com>.

39. General W Shelton, head of USAF Space Command, addressing the [US] Air Force Association's Global Warfare Symposium; reported by, J Tirpak, *Operationally responsive space not so responsive*, 19 November 2012, Air Force Magazine, Arlington, viewed 21 January 2013, <http://www.airforcemagazine.com>.

40. *'Starting with Columbia and continuing with Challenger, Discovery, Atlantis and Endeavour, the spacecraft has carried people into orbit repeatedly, launched, recovered and repaired satellites, conducted cutting edge research and built the largest structure in space, the International Space Station.' Space shuttle Endeavour attached to shuttle carrier aircraft*, 2012, National Aeronautics and Space Administration, Washington DC, viewed 15 September 2012, <http://www.nasa.gov>.

41. United States Congress, Office of Technology Assessment, *Anti-satellite weapons, countermeasures and arms control, op. cit.* p. 56. *SAINT* was also acknowledged by WWS Wong & J Fergusson, *Military space power, op. cit.* p. 6, citing B Rose, *Military space technology,* Midland Publishing, Surrey (UK), 2008. Also, by B Weeden, *China's BX-1 microsatellite: a litmus test for space weaponisation,* 20 October 2008, The Space Review, n.p., viewed 7 December 2012, <http://www.thespacereview.com>.

42. J Johnson-Freese, *Space as a strategic asset,* Columbia University Press, New York, 2007, pp. 138 – 139, citing M Bile, R Kane, & D Cox, *'Military microsats: matching requirements and technology',* paper presented to the AIAA space 2000 Conference and Exhibition, Long Beach California, 19 -21 September 2000.

43. J Johnson-Freese, *Space as a strategic asset, op. cit.* p. 189, citing EM Grossman & KJ Costa, 'Small experimental satellites may offer more than meets the eye,' *Inside the Pentagon*, 4 December 2003. A similar claim was recorded by; WWS Wong & J Fergusson, *Military space power, op. cit.* p. 91.

44. SM Pekkanen and P Kallender-Umezu, *In defense of Japan; from the market to the military in space policy,* Stanford University Press, Stanford, 2010, pp. 166 – 167.

45. J Johnson-Freese, *Space as a strategic asset, op. cit.* p. 224.

46. G Kulacki & D Wright, *A military intelligence failure?* 16 August 2004, Union of Concerned Scientists, Cambridge, viewed 13 March 2013, <http://www.ucsusa.org>.

47. B Weeden, *China's BX-1 microsatellite: a litmus test for space weaponisation, op.cit.*

48. MS Smith, *Surprise Chinese manoeuvres mystify Western experts,* 19 August 2013, Space and Technology Policy Group, Arlington, viewed 24 December 2013, <http://www.spacepolicyonline.com>._B Gertz, *China launches three ASAT satellites,* 26 August 2013, The Washington Free Beacon, Washington DC, viewed 24 December 2013, <http://freebeacon.com>.

49. *Kuaizhou challenges U.S. perceptions of Chinese military space strategy*, 27 September 2013, All Things Nuclear (sponsored by Union of Concerned Scientists, Cambridge),viewed 10 January 2014, <http://allthingsnuclear.org/kauizhou-challenges-u-s-perceptions-of-chinese-military-space-strategy/>. *Kuaizhou KZ-1 quick-response launch vehicle,* n.d., Global Security, Colorado, viewed 28 February 2014, <http://www.globalsecurity.org>.

50. *Electromagnetic railgun,* 2011, Office of Naval Research, Arlington, viewed 15 December 2011, <http://www.onr.navy.mil>.

51. IR McNab, 'Launch to space with an electromagnetic railgun,' *IEEE Transactions on Magnetics*, vol. 39, iss. 1, January 2003, p. 295 – 304. Located at IEEE Xplore, <http://ieeexplore.ieee.org>.

52. *Raytheon awarded naval power system contract*, 9 February 2012, Space War, Space Media Network, n.p., viewed 11 February 2012, <http://www.spacewar.com>.

53. DG Fearn, 'Application of ion thrusters to high-thrust, high-specific-impulse nuclear electric missions', C Bruno (ed.), *Nuclear space power and propulsion systems, op. cit.* p. 76.

54. J Ray, *One year later*, 14 August 2011, Spaceflight Now, n.p., (UK), viewed 27 August 2011, <http://www.spaceflightnow.com>.

55. *Northrop Grumman develops solar electric propulsion flight concepts for future space mission'*, 2 February 2012, Space Daily, Space Media Network, n.p., viewed 7 February 2012, <http://www.spacedaily.com>.

56. *Spacecraft propulsion: the future is electric*, December 2001, Snecma, Courcouronnes, viewed 15 September 2012, <http://www.snecma.com>. (The power was provided in kW without elaboration. Translation into Newton (metres per second^2) values was at the ratio 1W = 1N. Author.).

57. United States Congress, Office of Technology Assessment, *Anti-satellite weapons, countermeasures and arms control, op. cit.* p. 65.

58. Air University, Air Command and Staff College, *Space Handbook, A warfighter's guide to space*, Maxwell Air Force Base, Alabama, 1993, p. 30.

59. US Department of Defense, *Ballistic Missile Defense Review Report*, *op. cit.*

Chapter 6

1 Definitions. It is necessary to understand some terminology relating to treaties because those terms indicate the extent, if any, to which States consider themselves bound in law to adhere to treaty provisions. States commonly have options to be bound or not bound by treaties.

Treaties sponsored through the UN are usually opened for signature until a specified date. During that period, States may *sign* or become *signatories* to a treaty. This, broadly, is an interim commitment to not act against the intent of a treaty. States that are signatories then need to confirm their status as accepting the treaty. They do this by a further *ratification* process.

To *ratify* a treaty means the ratifying State agrees to be bound and to express that agreement in its domestic arrangements if a further process is required by domestic arrangements.

It is common to provide that States, which do not for any reason sign a treaty during the specified period may subsequently agree to be bound by the treaty through *acceding* to it. They normally perform this function by lodging an *instrument of accession*.

New members of the UN may, under a complicated web of arrangements, be assumed to have *succeeded* to a treaty. For example, a former colonial overlord might have ratified a treaty and the new State might be regarded as having, in effect, inherited the same obligation. The extent to which such States are thus bound is unclear and the related debate seems unlikely to be settled in the foreseeable future.

There exist variations to all of these concepts and they are considered in more detail by, *inter alia*, the *UN treaty handbook,* UN Reproduction Section, n.p., pp. 5 – 18.

Legal analysts have identified two broad categories of approaches adopted by States as they implement domestic obligations to comply with international agreements to which they have committed. *Monist* States tend to view international and domestic law as forming a unified whole, so the act of ratifying a treaty can incorporate it into domestic law. *Dualist* States perceive a distinction between international and domestic law. A treaty, for example, would need to be legislated in accordance with domestic constitutional processes or it would not form part of the domestic law. (MN Shaw, *International law*, 5th edn, Cambridge University Press, Cambridge, 2003, pp. 135 – 177.) The categorisation used by Shaw is not the only option. *Transformation* and *incorporation* are also employed. (Sir A Mason, 'International Law as a Source of Domestic Law', in BR Opeskin & DR Rothwell (eds), *International law and Australian federalism*, Melbourne University Press, Melbourne, 1997, p. 209.)

2 SC Neff, *War and the law of nations*, Cambridge University Press, Cambridge, 2005, p. 178.

3 Australia, for example, acquired materials for chemical warfare. (G Plunkett, *Chemical warfare in Australia: Australia's involvement in chemical warfare 1914 – 1945*, Australian Military History Publications, Loftus, 2007.) During 1930-31 Italian personnel crushed resistance in Libya, killing more than 50 000 civilians. The Italians used poisonous gas, blocked water wells, slaughtered herds of animals and participated in mass executions. Italy conducted a similar military campaign in Ethiopia during 1935 – 36. (S Morillo, J Black & P Lococo, *War in world history*, vol. 2. McGraw-Hill, New York, 2009, p. 525.)

4 *Summary of the Geneva Conventions of 12 August 1949 and their Additional Protocols*, 2012, 2nd ed. International Committee of the Red Cross, Geneva, viewed 22 July 2014, <http://www.icrc.org>.

5 S Metz, 'The future of the United Nations', HH Almond & JA Burger (eds), *The history and future of warfare,* Kluwer Law International, The Hague, 1999, p. 861.

6 Security Council. S/RES/188 (1964). *Complaint by Yemen.*

7 General Assembly. A/RES/2625 (XXV) (24 October 1970). *Declaration on the Principles of International Law Concerning Friendly Relations and Cooperation among States.*

8 HC von Sponeck, 'The United Nations and NATO', *Current Concerns, no. 2.* 2009, Current Concerns, Zurich, viewed 3 February 2014, <http://www.currentconcerns.ch/index.php?id=711>. (von Sponeck was a UN Assistant Secretary-General. Author) The existence of the document has been acknowledged by NATO. (*Five years of strengthened cooperation with the United Nations*, 23 September 2013, North Atlantic Treaty Organisation, Brussels, viewed 3 February 2014, <http://www.nato.int/cps/en/natolive/news_103374.htm?selectedLocale=en>.)

9 There is an example of the number of members of the Security Council having been increased. The number of non-permanent members was increased from six to ten. (General Assembly Resolution 1991 (XVIII) (17 December 1963) *Question of equitable representation on the Security*

Council and the Economic and Social Council.

10 General Assembly Resolution 2758 (XXV1) (25 October 1971), *Restoration of the lawful rights of the People's Republic of China to the United Nations.* (A notable but less fundamental change also occurred in 1991 when the USSR was replaced in the United Nations, including the Security Council, by Russia. Russia maintained responsibility for all rights and obligations for the USSR. A description of the related processes is at: YZ Blum, 'Russia takes over the Soviet Union's seat at United Nations', *European Journal of International Law*, vol. 3, 1992, pp. 354 – 361.)

11 The figures were: 1958 – 1967, 14 resolutions; 1968 – 1977, 16; 1978 – 1987, 26; 1988 – 1997, 23; and 1998 – 2007, 30. Subsequent to 2007 a further two resolutions were passed leading to a total of 111. (*Index of online General Assembly Resolutions relating to outer space*, 2014, United Nations Office for Outer Space Affairs, Vienna, viewed 29 January 2014, <http://www.oosa.unvienna.org>.)

12 *Status of international agreements relating to activities on outer space as* at 1 January 2013, Legal Subcommittee, United Nations Committee on the Peaceful Uses of Outer Space, Fifty-second session, 8 – 19 April 2013, reference A/AC.105/C.2/2013/CRP.5 dated 28 March 2013.

13 K-H Bockstiegel, 'Settlement of disputes regarding space activities', *Journal of Space Law*, vol. 21, no. 1, 1993, pp. 1 – 10.

14 M Listner, *Iridium 33 and Cosmos 2251 three years later*, 13 February 2012, The Space Review, n.p., viewed 29 January 2014, <http://www.thespacereview.com>.

15 LV Viikari, '*The environmental element in space law Assessing the present and charting the future*', Martinus Nijhoff, Leiden, 2008, pp. 74- 75.

16 M Hucteau, '*CNSS presentation space debris activities registration issues*' (Slide presentation), 2011, UN Office for Outer Space Affairs, Vienna, viewed 25 August 2011, <http://www.oosa.unvienna.org>.

17 *Registration of objects launched into outer space*, 2011, United Nations Office of Outer Space Affairs, Vienna, viewed 30 November 2012, < http://www.oosa.unvienna.org/oosa>.

18 J McDowell, *Adherence to the 1976 Convention on Registration of Objects Launched into Outer Space*, 2011, Space Report, n.p., viewed 30 November 2012, <http://planet4589.org>.

19 *Convention on Registration of Objects Launched into Outer Space,* New York, 14 January 1975, entry into force 15 September 1976.

20 WJ Broad, 'Earthlings at odds over Moon Treaty', *Science* (new series), vol. 206, no. 4421, (23 November 1979), pp. 915 – 916.

21 V Pop, 'Appropriation in outer space: the relationship between land ownership and sovereignty on the celestial bodies', *Space Policy*, vol. 16, iss. 4 (2000), pp. 275 – 282. LE Viikari, 'The legal regime for Moon resource utilisation and comparable solutions adopted for deep seabed activities', *Advances in Space Research*, vol. 31, iss. 11 (June 2003), pp. 2427 – 2432. H Caldicott & C Eisendrath, *War in heaven*, *op. cit.* p. 113.

22 General Assembly Resolution A/RES/49/34 (30 January 1995) *International cooperation in the peaceful uses of outer space, including the question of the review of the Agreement Governing the Activities of States on the Moon and other Celestial Bodies,* para 42.

23 A Gorman, 'Look, but don't touch', *The Conversation*, 29 November 2013, Melbourne, <http://www.noreply@theconversation.com>.

24 *A bill to establish the Apollo Lunar Landing Sites National Historical Park on the Moon, and for Other Purposes*, HR 2617, 113th Congress, 1st Session, 8 July 2013. As at June 2014 the title of the bill had been amended to, *Apollo Lunar Landing Legacy Ac*t and it had been referred to the (House) Subcommittee on Space. (Bill Summary and Status, 113th Congress (2013 -2014) H.T. 2617. Located at, <http://thomas.loc.gov/cgi-bin/bdquery/z?d113:h.r.02617:>.)

25 HR Hertzfeld & SN Pace, 'International cooperation on human lunar heritage', *Science*, vol. 342, 29 November 2013; located at. <http://www.sciencemag.org>.

26 In 1993 Russia through Sotheby's auctioned for USD 442 500 three particles of lunar regolith, weighing about 200 mg, that had been collected by a Soviet probe. The sale price equated to USD 2.2 million per gram. V Pop, 'The property status of lunar resources', V Badescu (ed.), *Moon prospective energy and material resources*, Springer, New York, 2012, p. 558.

27 *Law of the Sea Convention*, US Department of State, 2012, US Department of State, Washington DC, viewed 8 March 2012, <http://www.state.gov>.

28 *Hague Code of Conduct Against Ballistic Missile Proliferation*, The Hague, brought into effect 25 November 2002. Located at HCoC Executive Secretariat, <http://www.hcoc.at>.

29 There are minor discrepancies in figures given by various sources for the forms of accession, by State, to the five space-related agreements of treaty status. United Nations treaty records (https://treaties.un.org>) provided up-to-date (late January 2014) details of States parties to only the Moon Agreement and Registration Convention. Corresponding details for the remaining three arrangements were sourced from a document originated with the (UN) Committee on the Peaceful Uses of Outer Space, Legal Subcommittee, Fifty-second session, 8 – 19 April 2013, *Status of International agreements relating to activities in outer space as at 1 January 2013*, Legal Subcommittee, United Nations Committee on the Peaceful Uses of Outer Space, Fifty-second session, 8 – 19 April 2013, reference A/AC.105/C.2/2013/CRP.5 dated 28 March 2013.

30 N Mateesco Matte (ed.), *Space activities and emerging international law*, Center for Research of Air and Space Law, McGill University, Montreal, 1984.

31 A/AC. 105/935 (2009). *Report of the Legal Subcommittee on its forty-eighth session, held in Vienna from 23 March to 3 April 2009,* p. 10, paragraph 55 and, Annex II.

32 An Interagency Debris Coordination Committee, in 2002, published guidelines which led to a European Code signed by five European space agencies in 2006. (*Regulations and Treaties, Operations*, 30 November 2012, European Space Agency, Paris, viewed 8 January 2013, <http://

www.esa.int>.) For a brief history of such attempts, J Robinson, *Advancing an international space code of conduct,* 13 July 2012, e-International Relations, viewed 3 October 2012,<http://www.e-ir.info>.

33 United Nations General Assembly, resolution A/RES/65/68 (8 December 2010) *Transparency and confidence-building measures in outer space activities.*

34 United Nations Press Release DC/3441 OS/1820 of 18 July 2013, *UN group of governmental experts on transparency and confidence building measures in outer space concludes its work.*

35 FG Klotz, *America on ice,* National Defense University Press, Washington DC, 1990, pp. 116 – 128.

36 D Leary, *'Briefing note: the Australian Federal Court Judgement in the Japanese Whaling Case Humane Society International Inc v. Kyodo Senpaku Kaisha Ltd'.* Shown as [2008] FCA 31, but actually [2008] FCA 3. Located at <http://www.indymedia.org.au/2010/01/10/legal-cases-rulings-re-whaling-in-australian-southern-ocean-2008>. JD Myhre, *The Antarctic Treaty system: politics, law and diplomacy,* Westview Press, Boulder, 1986, p. 14. '*The Australian claim* [to Antarctica from 160°E to 45°E] *is relatively solid, based on an effective occupation following acquisition of inchoate title conferred by discovery and exploration. It is weakened, though, by its vast size, covering 115° out of 360°...*'

37 A Darby, *China flags its Antarctic intent,* 11 January 2010, Sydney Morning Herald, Sydney, viewed 21 December 2012, <http://www.smh.com.au>.

38 The PRC had notified 24 placements of bases and instrumentation, India 5 and Russia 6. (Electronic Information Exchange System, 2011, Secretariat of the Antarctic Treaty, <http://www.ats.aq>.)

39 *Mineral activities and environmental issues in Antarctica,* 2012, Gateway Antarctica, University of Canterbury, Christchurch, viewed 28 April 2012, <http://www.anta.canterbury.ac.nz>. *Explanatory notes for the mineral resources map of the Circum-Pacific region Antarctic sheet,* 1998, US Geological Survey, US Department of the Interior, Reston, viewed 6 May 2012. Located at <http://pubs.usgs.go.edu.auv>.

Chapter 7

1 References are:

K Marx & F Engels, *Selected works,* vol. 1, Progress Publishers, Moscow, 1969, pp. 38 – 50.

EH Carr, *The twenty years' crisis,* 2nd ed. (M Cox), Palgrave, New York, 2001, (Originally published 1939).

HJ Morgenthau, *Politics among nations,* 6th edn, Alfred Knopf, New York, 1985. (Originally published 1948, 6th edn developed by KW Thompson [Morgenthau, d. 1980]).

RO Keohane & JS Nye, *Power and independence,* Little Brown, Boston, 1977.

A Linklatter, *Men and citizens in the theory of international relations*, 2nd edn, Macmillan Press, 1990.

B Buzan, *People, states and fear*, 2nd edn, Harvester Wheatsheaf, Hemel Hempstead, 1991.

N Chomsky, *World orders old and new*, Columbia Press, New York, 1996, p. 120 *et ceq.*

2 H Kissinger, *Does America need a foreign policy*? Simon and Schuster, New York, 2001.

3 S Laycock, *All the countries we've ever invaded and the few we never got round to*, History Press, Gloucestershire, 2012.

4 Background Note: *Canada*, 2012, US Department of State, Washington DC, viewed 21 April 2012, <http://www.state.gov>. *So, how's it going, eh?,* 21 February 2006, CBC/Radio, Ontario, viewed 22 April 2012, <http://www.cbc.ca>.

5 References are:

Member States of the United Nations, n.d., United Nations Homepage, New York, viewed 29 January 2014, <http://www.un.org>.

United Nations Committee on the Peaceful Uses of Outer Space, *Status of International agreements relating to activities in outer space as at 1 January 2013*, Legal Subcommittee, Fifty-second session, 8 – 19 April 2013, *op. cit.*

Registration Convention, *Convention on registration of objects launched into outer space, status as at 28 January 2014,* United Nations Treaty Collection, Databases, Geneva, viewed 29 January 2014, <https://treaties.un.org/pages/ViewDetailsIII.aspx?&mtdsg_no=XXIV~1&chapter=24&Temp=mtdsg3&lang=en>.

Moon Agreement, *Agreement governing the activities of States on the Moon and other celestial bodies, status as at 29 January 2014,* United Nations Treaty Collection, Databases, Geneva, viewed 29 January 2014, <https://treaties.un.org/pages/ViewDetailsIII.aspx?src=IND&&mtdsg_no=XXIV-2&chapter=24&lang=en>.

NPT, *Treaty on the Non-Proliferation of Nuclear Weapons*, 1 July 1968, entry into force, 5 March 1970.

MTCR partners, October 2013, Missile Technology Control Regime, n.p., viewed 29 January 2014, <http://www.mtcr.info>.

Comprehensive TBT, *Current treaty status*, 8 March 2013, Comprehensive Test Ban Treaty Organisation, Vienna, viewed 29 January 2014, <http://www.ctbto.org

Hague Code of Conduct Against Ballistic Missile Proliferation, *The Hague Code of Conduct Against Ballistic Missile Proliferation, Subscribing States*, 19 August 2013, HCoC, (hosted by) Austrian Foreign Ministry, Vienna, viewed 29 January 2014, <http://www.hcoc.at>.

Biological (and Toxic) Weapons Convention, *Membership of the biological weapons convention, Membership, States Parties,* c. 2013, United Nations Office at Geneva, viewed 29 January 2014, <http://www.unog.ch>.

Chemical Weapons Convention, *OPCW Member States, Non-Member States*, c. March 2013, Organisation for the Prohibition of Chemical Weapons, The Hague, viewed 29 January 2014, <http://www.opcw.org>.

Conventional Weapons, *The Convention on Certain Conventional Weapons,* United Nations Office at Geneva, viewed 29 January 2014, <http://www.unog.ch>.

Landmines Convention, *Anti-Personnel Landmines Convention, States parties and signatories*, n.d., United Nations Office at Geneva, viewed 30 January 2014, <http://www.unog.ch>.

Convention on Cluster Munitions, *Cluster munitions, signatories and ratifying states,* n.d., United Nations Office for Disarmament Affairs, Geneva, viewed 30 January 2014, <http://www.un.org/disarmament>.

Rome Statute of the International Criminal Court, *The Rome Statute in the world*, 10 November 2011, Coalition for the International Criminal Court, The Hague, viewed 30 January 2014, <http://www.iccnow.org>.

6 Signed by President Clinton, but US support for the Statute was withdrawn under the subsequent Bush administration. For a summary of events; Bush approach to the ICC, c.2002, American Non-Governmental Organizations Coalition for the International Criminal Court, New York, viewed 23 March 2013, <http://www.amicc.org>.

7 T Shinabu, 'China's bilateral treaties 1973 – 1982: a quantitative study,' International Studies Quarterly, vol. 31, no. 4, 1987, pp. 439 – 456, p. 439. Treaties in Force 2013, Department of State, Bureaus/Offices Reporting Directly to the Secretary, Office of the Legal Advisor, Treaty Affairs, Washington DC, <http://www.state.gov/s/l/treaty/tif/index.htm>.

8 For example, Russia has raised concerns that newly-developed weapons have almost erased the distinction between nuclear [presumably tactical] weapons and conventional weapons. (*Russia calls on NATO to review Cold War methods of arms control*, 31 July 2013, Space War, Space Media Network, n.p., viewed 24 December 2013, <http://www.spacewar.com>.)

9 *Monroe Doctrine; December 2, 1823*, 2008, Lillian Goldman Law Library, Yale Law School, New Haven, viewed 23 February 2012, <http://www.avalon.law.yale.edu>. (*United States policy on the new political order developing in the rest of the Americas and the role of Europe in the Western Hemisphere*, address to Congress by President J Monroe, 2 December 1823.) R Henig, *Versailles and after 1919 – 1933*, 2nd edn, Routledge. London, 1995, p. 16.

10 Proclamation No 2667, *Policy of the United States with Respect to the Natural Resources of the Subsoil and Seabed of the Continental Shelf*, 28 September 1945. (Proclamation No 2667), 10 FR 12303 (1945).

11 *Project Mercury, manned space flight (MA-9)*, National Security Action Memorandum 3 May 1963. (declassified 23 November 1963)

12 *Use of airspace by US military aircraft and firings over the high seas,* US Department of Defense Directive 4540.1, January 13, 1981 (Current as of December 8, 2003), paragraphs 5.1 and 5.2.)

13 *Overseas US territories and associated States*, 7 January 2014, US Government Official Web Portal, Washington DC, viewed 30 January 2014, <http://www.usa.gov>.

14 Europe 350 sites, East Asia and Pacific 175 sites, North Africa Near East and South Asia 13 sites, Western Hemisphere 6 sites and Sub-Saharan Africa 2 sites. (V Reed, *US military bases in foreign nations; a summary of the Pentagon's data*, 2007, Straus Military Reform Project, Centre for Defense Information, Washington DC, viewed 23 February 2012, <http://www.cdi.org>.) However, the number of overseas US military bases is a topic of debate with claims the 'real' total is between 700 and 1 000. (*America's empire of bases. How many military bases does the US really have around the globe?* 20 January 2011, The Nation, New York, viewed 30 January 2014, <http:///www.thenation.com/article/157595/americas-empire-bases-20#>.)

15 Background Note*: China*, 6 September 2011, US Department of State, Washington DC, viewed 17 May 2012, <http://www.state.gov>. *Westinghouse AP1000 Nuclear Plants, China*, c.2010, power-technology.com, London, viewed 30 January 2014, <http://www.power-technology.com/projects/westinghouseap1000/>. These exports from the US continued through to and including 2013. (Chart – *China – Imports – Evolution NCE: Nuclear reactors, boilers, machinery and mechanical appliances; parts thereof – Annual FOB USD*, 2014, Business Research Service Trade, international, viewed 30 January 2014, <http://trade.nosis.com/en/Comex/Import-Export/China/Nuclear-reactors-boilers-machinery-and-mechanical-appliances-parts-thereof/CN/84>.)

16 *United States Air Force chronology*, 3 December 2008, Hill Air Force Base, Utah, viewed 25 September 2013, <http://www.hill.af.mil>, pp. 31 – 32.

17 National Security Council, NSC 5520, *Draft statement of policy on US scientific satellite program*, 20 May 1955. Located at Internet Archive, n.p., (US) <http://archive.org>.

18 Note by the Executive Secretary to the National Security Council, NSC 5814, 20 June 1958, *US policy on outer space*, para 43. c. (declassified from SECRET, date of declassification illegible). Located at; US Department of State, <http://history.state.gov/historicaldocuments>

19 Note by the Executive Secretary to the National Security Council, NSC 5918, *US policy on outer space,* 17 December 1959, para 19 (declassified from SECRET, 13 August 1991). References to a related document dated January 1960 were found but it was not located. In most sources consulted it was either omitted or had been deleted.

20 *National Space Policy of the United States of America*, President of the United States, 28 June 2010, (heading) 'Space Nuclear Power'.

21 *International Arms Regulations.* (In force, date uncertain.) (These are the conditions for the import and export of items and services identified by a United States Munition List which in turn implements the requirements of 22 U.S.C. 2778 of the Arms Control Export Act.)

22 D Messier, *Will a new space power rise along the Atlantic?*, 15 August 2011, The Space Review, n.p., viewed 13 June 2011, <http://www.thespacereview.com>.

23 Y Zaitsev, *Russia begins elbowing Ukraine out from Brazil's space program, op. cit.* D Messier, *Brazil scales back launch vehicle plans, op. cit.* A Svitak, *Ukraine, Brazil prepare for 2015 Cyclone 4 launch,* n.d., Aviation Week & Space Technology, New York, viewed 3 January 2014, http://www.aviationweek.com

24 Ministry of Defense [Brazil], *National strategy of defense*, 1st edn, decree no. 6703, 18 December 2008, p. 32.

25 A e Silva, F Liporace & M Santos, *CBERS: A Reference in the Brazilian Space Program*, 2007. Located at US Geological Survey, Washington DC, viewed 21 February 2012, <http://www.calval.cr.usgs.gov>.

26 *Comtech to provide satellite system for Brazilian military*, 12 January 2014, Space Newsfeed, Nairn, viewed 14 January 2014, <mikeas37@spacenewsfeed.co.uk>. (Site scheduled to move to <www.spacenewsfeed.com>)

27 *Find an embassy*, 2012, Foreign and Commonwealth Office, London, viewed 12 July 2012, <http://www.fco.gov.uk>.

28 *Joint fact sheet: US and UK Defense Cooperation*, 14 March 2012, Office of the President of the United States, Washington DC, viewed 12 July 2012 <http://www.whitehouse.gov/administration>. *UK-US economic relations*, n.d., Foreign and Commonwealth Office, British Embassy Washington, and *US–UK* [military] *relations*, 2012, Foreign and Commonwealth Office, London, viewed 3 June 2012,<http://www.fco.gov.uk>.

29 R Norton-Taylor, *Britain's nuclear arsenal is 225 warheads, op. cit.*

30 *Military uses of space*, Postnote Number 273, December 2006. Located at Parliamentary Office of Science and Technology, London, viewed 12 July 2012, <http://www.parliament.uk. T Dinerman, *Future British military space policy*, 12 April 2004, The Space Review, n.p., viewed 14 July 2012, <http://www.thespacereview.com>.

31 *UK space capability development,* n.d., speech by Air Vice-Marshal B North, Assistant Chief of the Air Staff, Royal Air Force, Ministry of Defence, London, <http://www.raf.mod.uk>.

32 For aspects of the encirclement:

US Department of Defense, *Ballistic Missile Defense Review Report, op. cit.* pp. 15, 24, 29 –33.

E Tauscher, (US Under Secretary for Arms Control and International Security) *address to 12th Royal United Services Institute (RUSI) missile defense conference,* 16 June 2011, US Department of State. Located at US Department of State, Washington DC, viewed 4 July 2011, <http://www.state.gov>.

Floating X-band radar returns to Pearl Harbour, 20 October 2011, Global Security Newswire, n.p., viewed 1 November 2011, <http://gsn.nti.org>.

Opposition deputies [in Turkey] *protest against NATO radar system*, 10 March 2012, Space War, Space Media Network, n.p., viewed 19 March 2012, <http://www.spacewar.com>.

For the US official attitude; HR Clinton, *Remarks at the US Institute of Peace, China Conference*, 7 March 2012, US Department of State, viewed 15 March 2012, <http://www.state.gov>, during which Secretary of State Clinton was recorded as stating: *'Our economies are tightly entwined. And so is our security. We face shared threats like nuclear proliferation, piracy, and climate change, and we need each other to solve these problems. The opportunities before us are also shared, and they define our relationship more than the threats.'*

P Hartcher, *US Marine base for Darwin,* 11 November 2013, The Sydney Morning Herald, Fairfax Media, Sydney, viewed 7 March 2014, <http://www.smh.com.au>.

33 *China's Space Activities*, (White Paper), 15 December 2003, China National Space Administration, n.p., <http://www.cnsa.gov.cn>.

34 *China's Space Activities*, (White Paper), *op. cit.* Aims and Principles;

35 *China's Space Activities in 2011*, (trans, no name), Information Office of the State Council, Beijing, viewed December 2011, <http://www.linkedin.com>.

36 B Harvey, *China's Space Program*, Praxis Publishing, Chichester, 2004.

37 *China needs its own space laws,* 13 March 2012, Space Daily, Space Media Network, n.p., <http://www.spacedaily.com>..

38 JC Moltz, *The politics of space security: strategic restraint and the pursuit of national interests,* Stanford University Press, Stanford, 2008, p. 289.

39 *Korea North*, 28 January 2014, The World Factbook, Central Intelligence Agency, n.p., (US), viewed 8 July 2014, <https://www.cia.gov/library/. (Per capita USD 1 800 for 2011 noted as having been estimated from unreliable data.) *Korea South*, 20 June 2014, The World Factbook, *op. cit.* (Per capita USD 33 200, 2012 estimated)

40 *China, Russia discuss N. Korea nuclear tes*t, 24 February 2013, Global Times, Beijing, viewed 24 February 2013, <http://www.globaltimes.cn>.

41 *Security Council demands that DPR Korea suspend ballistic missile activities*, 15 July 2006, UN News Centre, New York, viewed 23 January 2014, <http://www.un.org/News>. (For example; Security Council Resolution 1718, adopted at meeting 5551 on 14 October 2006, reference S/RES/1718 (2006).) Security Council. S/RES/1695 (2006). *Letter dated 4 July 2006 from the Permanent Representative of Japan to the United Nations addressed to the President of the Security Council* (S/2006/481).

42 *Outcry over N Korea missile*, 5 July 2006, BBC News, London, viewed 19 May 2012, <http://news.bbc.co.uk>. B Harden, *Defiant N Korea launches missile*, 5 April 2009, The Washington Post, Washington DC, viewed 19 May 2012, <http://www.washingtonpost.com>. T Schwartz, *North Korea rocket breaks up in flight*, 17 April 2012, Cable News Network, n.p., viewed 19 May 2012, <http://www.cnn.org >.

There was controversy over whether the rocket carried a satellite, whether the satellite had reached orbit and whether the DPRK attempted to notify the UN as required by the Registration Convention. (*North Korea didn't notify UN of satellite*, 9 April 2009, Thomson Reuters, n.p., viewed 26 February 2012, <http://www.reuters.com >.)

43 *North Korea successfully launches long-range rocket*, 12 December 2012, Space War, Space Media Network, n.p., viewed 16 December 2012, <http://www.spacewar.com>. *N. Korea's satellite 'orbiting normally'*, 13 December 2012, Space War, Space Media Network, n.p., viewed 16 December 2012, <http://www.spacewar.com>.

44 The official DPRK website included the following description of Kim Jong Il climbing to and reaching the peak of Mount Paektu during a blizzard:

At that moment there was a sudden thunderous sound as if the ice was breaking on Lake Chon, and the furious blizzard stopped blowing. The clouds, which were floating on the ridge, gradually moved to one side, the sun shone and a majestic snowscape was presented; the ridges of the mountain covered with snow were glistening. Impressed by the mystery, the entourage exclaimed. With a smile on his face, he [Kim Jong Il] *said it seemed that Mt Paektu knew its master.*

DPR Korea Politics, Democratic People's Republic of Korea, n.p., viewed 25 February 2012, <http://www.korea-dpr.com>.

45 SP Cohen, 'India, Pakistan and Kashmir,' in S Ganguly(ed.), *India as an emerging power*, Frank Cass, London, 2003, pp. 45 – 56. D McLeod, *India and Pakistan: friends, rivals or enemies?* Ashgate Publishing, Aldershot, 2008, p. 1. '*Since the partition of India and Pakistan in 1947 the relationship between these two states has been the most intractable and the most dangerous political stand-off in South Asia.*'

46 A Pasricha, *India, China work to resolve border dispute,* 20 October 2013, Voice of America , Washington DC, viewed 21 October 2013, <http://www.voanews.com>.

47 N Mohanty, *America, Pakistan and the India factor*, Palgrave Macmillan, New York, 2013, pp. 67 – 141. P Mishra, 'Dear Uncle Sam..why do India and Pakistan see America in such opposite ways?' *Foreign Policy*, iss. 188, September/October 2011 pp. 104 - 107. H Kissinger, *Does America need a foreign policy? op. cit.* pp. 153 – 160.

48 *Countering China, Obama backs India for U.N. Council*, 8 November 2010, The New York Times, New York, viewed 19 May 2012, <http://atwar.blogs.nytimes.com>.

49 *Indian Space Research Organisation, 2010-2011 Annual Report*, Indian Space Research Organisation, Bangalore, viewed 27 January 2012, <http://www.isro.org>.

50 AJ Tellis, 'Toward a 'force-in-being'; the logic, structure and utility of India's emerging nuclear posture,' S Ganguly (ed.), *India as an emerging power, op. cit.* pp. 61 - 108.

51 TR Mattair, *Global security watch – Iran: a reference handbook*, Praeger security International, Westport, 2008, p. 19.

52 *Iran defends development of advanced centrifuges*, 12 January 2014, Space War, Space Media Network, n.p., viewed 14 January 2014, <http://www.spacewar.com>.

53 *About Iranian Space Agency,* 2012, Iranian Space Agency, Presidency of the Islamic Republic of Iran, Tehran, viewed 28 February 2012, <http://www.isa.ir>. Also, SK Dehgan, *Iran launches monkey into space,* 28 January 2013, The Guardian, London, viewed 5 February 2013, <http://www.guardian.co.uk >. Iran has apparently launched into space one rat, one turtle, several worms and two monkeys (one of which was associated with a launch where its '*anticipated objectives were not accomplished*'). (*Iran admits monkey launch a failure,* 13 October 2011, Australian Broadcasting Commission, Sydney, viewed 13 January 2013, <http://www.abc.net.au/news/2011-10-13/iran-admits-failure-to-launch monkeys-into space/3568954>.)

54 PR Kumaraswamy, *Israel-China arms trade,* 16 July 2012, Middle East Institute, Washington DC, viewed 15 October 2012, <http://www.mei.edu>.

55 B Opall-Rome, *Israeli official urges space-based weapons,* 11 January 2005, Defense News, n.p., viewed 20 January 2005, p. 12. Located at *March 2005 Bibliography,* Arms Control Today <http://www.armscontrol.org>.

56 For example, A Di Filippo, 'Out of kilter: Japan's flawed nuclear disarmament approach and the NPT,' c.2006, *The Asia-Pacific Journal*; located at Japan Focus, Tokyo, <http://www.japanfocus.org>.

57 *Basic plan for space policy,* June 2009, Secretariat of Strategic Headquarters for Space Policy, Tokyo. Located at; Prime Minister of Japan and His Cabinet, Tokyo, viewed 6 February 2012, <http://www.kantei.go.jp>.

58 SM Pekkanen & P Kallender-Umezu, *In defense of Japan, op. cit.* pp. 95 -129. *H-IIA Launch Services,* n.d., Mitsubishi Heavy Industries, Tokyo, viewed 15 October 2012, <http://www.mhi.co.jp>.

59 *Japanese Aegis destroyer intercepts ballistic missile target,* October 2009, Military Photos, n.p., viewed 29 February 2012, <http://www.militaryphotos.net>. (This followed a similar achievement by the US Navy; *Navy says missile smashed wayward satellite,* 21 February 2008, The Associated Press, New York, viewed 29 February 2012, <http://www.msnbc.msn.com >.)

60 A Sattar, *Pakistan's foreign policy, 1947 – 2005: a concise history,* Oxford University Press, 2007, pp. 50 and 57 – 58; for the related Pakistan-India War, pp. 88 -104.

61 Inter Islamic Network on Space Sciences and Technology, Karachi, viewed 1 March 2012, <http://www.isnet.org.pk>.

62 *The aftermath of the Bolshevik revolution,* 2009, The History Guide, n.p., viewed 24 February 2011, <http://www.historyguide.org>. BD Rhodes, *The Anglo-American winter war with Russia: a diplomatic and military tragicomedy, 1918 – 1919,* Greenwood Press, New York, 1988.

63 *National security concept of the Russian Federation,* 10 January 2000, The Ministry of Foreign Affairs of Russia, Moscow, viewed 12 March 2012, <http://www.mid.ru>.

64 *Status and functions,* n.d., Federal Space Agency (Roscosmos), Moscow, viewed 12 March 2012, <http://www.federalspace.ru>.

65 Source of NATO member states; *NATO member countries,* 9 April 2013, North Atlantic Treaty

Organisation, Brussels, viewed 1 February 2014, <http://www.nato.int>. Source of EU member States; *Member States of the EU,* 2013, European Union, viewed 1 February 2014, <http://europa.eu>. Source of Eurozone members; *Eurozone group, members,* 2013, Eurozone Portal, n.p., (Europe), viewed 1 February 2014, <http://www.eurozone.europa.eu>. Source of ESA membership; *New member States, European Space Agency,* 2013, European Space Agency, Paris, viewed 1 February 2014, <http://www.esa.int/About_Us/Welcome_to_ESA/New_member_States>.

66 T Shanker & DM Herszenhorn, *US official says missile-defense shield will move forward,* 2 December 2011, The New York Times, New York, viewed 10 October 2011, <http://www.nytimes.com>.

67 The US has announced the missile defence arrangements for Europe will include upgraded versions of the SM-3 missile, a weapon with anti-satellite capabilities. (Office of the [White House] Press Secretary, *Fact sheet on US missile defense policy,* 17 September 2009, The White House, Washington DC.)

68 *European space policy,* 2011, European Commission, Brussels, viewed 13 March 2012, <http://ec.europa.eu>.

Chapter 8

1 Zhang Ju'nan, 'Fundamental ways to ensure space security', *Celebrating the space age,* United Nations Institute for Disarmament Research, United Nations, New York, 2007, pp. 109 – 111.

2 S Foerster, 'Strategies of Deterrence', in S Jasper (ed.), *Conflict and cooperation in the global commons, op. cit.* pp. 64 -66. J Child & D Faulkner, *Strategies of cooperation,* Oxford University Press, 1998.

3 *Global arms sales 2004 to 2011,* (Chart) 2013, Global issues, viewed 28 January 2012, <http://www.globalissues.org>. RF Grimmett & PK Kerr, *Conventional arms transfers to developing nations, 2004 – 2011,* Congressional Research Service, Washington DC, 2012, (Figure 2. Arms transfer agreements worldwide (supplier percentage of value); and Figure 3. Arms transfer agreements with developing nations (supplier percentage of value). Located at Federation of American Scientists, <http://www.fas.org>.

4 *Nuclear-free legislation - nuclear-free New Zealand,* New Zealand History online, Ministry for Culture and Heritage, Wellington, <http://www.nzhistory.net.nz>. Text of ANZUS Treaty, 1951 ANZUS Treaty, vol. 17, Historical Documents of Australia, Department of Foreign Affairs and Trade, <http://www.dfat.gov.au>.

5 *National space policy of the United States of America,* President of the United States, 28 June 2010, 'Preserving the Space Environment' (heading). *AU space primer,* 2003, Air University, National Space Studies Center, Maxwell, Alabama, <http://space.au.af.mil>, Ch. 5, Space law, policy and doctrine, p. 5-2 – 5-25. Located at; National Space Studies Center, Air University, Maxwell, <http://space.au.af.mil>. (*'The US adheres to the premise that any act not specifically prohibited is permissible.' 'Another widely accepted premise is that treaties usually regulate activities between signatories only during peacetime.'*)

6 *China's space activities in 2012* (White Paper), heading 'Purposes and principles of development', 2012. Located at China Daily, n.p., (PRC), viewed 31 October 2012, <http://www.chinadaily.com.cn>.

7 *Law of the Russian Federation about space activity*, Decree No. 5663-1 of the Russian House of Soviets, dated 20 August 1993. Article 3 (elaborated by Article 7). Located at; United Nations Office for Outer Space Affairs, <http://www.oosa.org>.

8 *National security space strategy* (Unclassified Summary), 2011, *op. cit.* p. 10, 11.

9 J Benitez, *US seeking wargames and combined operations with allies in space*, 13 June 2012, NATO Source, Atlantic Council, Washington DC, viewed 31 October 2012, <http://www.acus.org>.

10 G Apalin and U Mityayer, *Militarism in Peking's policies,* trans. P Doria, Progress Publishers, Moscow, 1980. BW MacDonald, China, Space Weapons and US Security, Council on Foreign Relations, New York, 2008. *Russian space defense troops launched*, 1 December 2011, *op. cit.*

11 Figure courtesy of JA Kennewell and B-N Vo, *An overview of space situational awareness*, 2013, authors' publication, <http://ba-ngu.vo-au.com/vo/KV_SSA_FUSION13.pdf>.

12 References are:

US national space policy, under the heading, 'Preserve the space environment', includes; '*Develop, maintain and use space situational awareness (SSA) information from commercial, civil and national security sources to detect, identify, and attribute actions in space that are contrary to responsible use and the long-term sustainability of the space environment.*' (p. 7) Also, under the heading, 'National security space guidelines', '*Maintain and integrate space surveillance, intelligence and other information to develop accurate and timely SSA. SSA information shall be used to support national and homeland security, civil space agencies, particularly human space flight activities, and commercial and foreign space operations;*' (pp. 13 -14). (National space policy of the United States of America, 28 June 2010, *op. cit.*)

Britain maintains data and monitors the activities of foreign satellites. (Speech by Air Vice-Marshal B North, Assistant Chief of the Air Staff, *UK space capability development, op. cit.*

Australia. '*Australia and the United States have agreed in a formal statement that 'SSA can be used to identify the behaviours of space assets, such as changes to courses or orbits. This can be used to determine if actions in space, such as collisions, are deliberate or accidental.*' (Australia-United States Ministerial Consultations, Australia-United States space situational awareness partnership fact sheet, 2010. Located at; Department of Foreign Affairs and Trade <http://www.dfat.gov.au>.)

13 W Kingston & K Scally, *Patents and the measure of international competitiveness*, Edward Elgar, Cheltenham, 2006. A Gibbs & B DeMatteis, *Essentials of patents*, John Wiley, Hoboken, 2003. A patent filing in the US was located referring to a modified form of airborne laser device. (United States Patent 7821623; Inventor DA Galasso; filing date 16 November 2004, publication date 26 October 2010. (The data of publication coincided approximately with the time at which the US Airborne (COIL) Laser project was cancelled.)

14 US Departments of Defense and State, *Risk assessment of United States space export control policy,*

op. cit. Proliferation of space-related space technologies has equipped several States with launch technology and many more with the capacity to build indigenous satellites as well as some with space-related weapons. (B Chapman, *Space warfare and defense, op. cit.* pp. 183 -222. *US, Thales at odds over request for ITAR-free satellite design*, 6 January 2012, Space News, Space Media Network, n.p., viewed 30 August 2012, <http://www.spacenews.com>.) (Thales had claimed to have designed and built a satellite not incorporating any ITAR-constrained components and therefore claimed also a right to sell the satellite to the PRC.)

15 *National Security Space Strategy* (Unclassified Summary), January 2011, Department of Defense and Office of National Intelligence, Washington DC, <http://www.defense.gov>, p. 2, Graph, 'Number of nations and government consortia operating in space'.

16 W Reynish, *The real reason selective availability* [of GPS] *was turned off* [during 1 May 2000], 1 July 2000, Avionics Today, Aviation Today, n.p., viewed 3 March 2013, <http://www.aviationtoday.com>. RA Williamson, HR Hertzfeld & J Cordes, *The socio-economic value of improved weather and climate information*, 2003, Space Policy Institute, George Washington University, Washington DC, viewed 3 March 2013, <http://www.gwu.edu>.

17 C Sanger, *Ordering the oceans: the making of the law of the sea*, University of Toronto Press, Toronto, 1987, pp. 62 – 64, 70 – 77.

18 RC Harding, *Space policy in developing countries, op. cit.* pp. 191 – 193.

19 HR Hertzfeld & SN Pace, 'International cooperation on human lunar heritage', *Science*, vol. 342, 29 November 2013; located at; <http://www.sciencemag.org>.

20 N Mateesco Matte (ed.), *Space activities and emerging international law, op. cit.* DA Ross, 'Canada's international security strategy, Beyond reason but not hope?' *International Journal*, vol. 65, Spring 2010, pp. 349 – 361.

21 The figures were: 1958 – 1967, 14 resolutions; 1968 – 1977, 16; 1978 – 1987, 26; 1988 – 1997, 23; and 1998 – 2007, 30. Subsequent to 2007 additional resolutions were passed leading to a total of 111. Located at UN Office for Outer Space Affairs 2011, UN, Vienna, viewed 25 January 2011, <http://www.oosa.unvienna.org>.

22 *History of US votes on UN space treaty*, December 2009, Global Network Against Weapons and Nuclear Power in Space, Brunswick, http://www.space4peace.org.

23 On 18 December 2012 the General Assembly adopted without vote a resolution on international cooperation in the peaceful uses of outer space. It was a non-binding expression of concern relating to the lack of progress made in relation to the prevention of an arms race in outer space. *Resolution adopted by the General Assembly, 67/113, International cooperation in the peaceful uses of outer space*, 18 December 2012 (distributed 14 January 2013), A/RES/67/113.

24 Resolution adopted by the General Assembly on 8 December 2010, *Transparency and confidence-building measures in outer space activities*, A/RES/65/68 (8 December 2010).

25 J Hermida, *Legal basis for national space legislation*, Kluwer Academic, Dordrecht, 2004.

26 For example, the US Deputy Assistant Secretary, Bureau of Arms Control, Verification and compliance, FA Rose, addressing the installation of missiles in NATO States (the European Phased Adaptive Approach), stated; '*In implementing the approach in Europe, we designed the EPAA to protect our deployed forces and Allies in Europe, as well as improving protection of the US homeland against potential ICBMs from the Middle East.*' (FA Rose, *Growing global cooperation on ballistic missile defense*, 10 September 2012, US Department of State, viewed 9 January 2013, <http://www.state.gov>.

27 *Persian Gulf states speed up US missile shield*, 1 October 2012, United Press International, Washington DC, viewed 13 December 2012, <http://www.upi.com>.

28 T Shinabu, 'China's bilateral treaties, 1973 – 1982' *op. cit.* p. 439.

29 *Treaties in force*, 1 January 2013, US Department of State, *op. cit.*

30 S Latchford, '*Strategies for defeating commercial imagery systems*', 2005, Occasional Paper No 39, Center for Strategy and Technology, Air University, Air University, Maxwell, viewed 9 January 2013, <http://www.au.af.mil>. Also, DL Sleeth, *Commercial imagery satellite threat*, 2004, Joint Military Operations Department, Naval Warfare College; located at; Defense Technology Information Center, Fort Belvoir, <http://www.dtic.mil>. F Fassihi & P Sonne, *A top satellite provider cuts off Iran state broadcaster*, 15 October 2012, The Wall Street Journal, New York, viewed 9 January 2013, <http://online.wsj.com>.

31 For example, Australia purchased part of the communications package on the Optus C1 satellite in 2003 (*Optus C1 Satellite Fleet*, Optus Singtel Australia, n.p., viewed 9 January 2013, <http://www.optus.com.au>.) Australia also purchased one satellite in a State (US)-owned six satellite constellation. (*Australia to join with United States in defence global satellite communications capability*, Australian Minister for Defence, 3 October 2007, Department of Defence, Canberra, viewed 9 January 2013, <http://www.defence.gov.au>.)

32 *Satellite shopping in China*, c. 2008, Telesatellite Global, n.p., viewed 9 January 2013, <http://www.tele-satellite-global.com>. Also, *Russian Satellite Communications Company*, 2012, Russian Satellite Communications Company, Moscow, viewed 9 January 2013, <http://eng.rscc.ru>.

33 B Roberts, 'Technology diffusion and international security,' J Brown (ed.), *Arms control in a multi-polar world*, VU University Press, Amsterdam, 1996.

34 Data. Table: *Merchandise trade (% of GDP)*, The World Bank, 2012, Washington DC, viewed 6 November 2012, <http://data.worldbank.org>.

35 MA Lorell, JF Lowell, RM Moore, *et al. Going Global? U.S. Government policy and the defense aerospace industry*, RAND Corporation, Santa Monica, 2002. *Aerospace Global Report 2011*, Clearwater Corporate Finance, London, viewed 6 November 2012, <http://www.clearwatercf.com>. *EADS Astrium to help Kazakhstan become competitive player in new satellites integration facility*, 22 May 2009, Satellite Today, Rockville, viewed 6 November 2012, <http://www.satellitetoday.com>.

36 Space Foundation, *The space report 2013*, p. 5, Exhibit 2. Space foundation index vs. other market indexes, 2013, The Space Foundation, Colorado Springs, viewed 20 February 2014, <http://www.spacefoundation.org>.

37 LI Sennott, *Overlapping false alarms*, c. 1987, Electrical Engineering, Stanford University, Stanford, viewed 5 November 2012, <http://www-ee.stanford.edu>. G Forden, P Podvig & TA Postol, *False alarm, nuclear danger*, IEEE Spectrum, vol. 37, no 3, March 2000. Located at; Center for Arms Control, Energy and Environmental Studies, Moscow, <http://www.armscontrol.ru>.

38 G Forden, T Postel & P Podvig, *False alarm, op. cit.* A Carter, *How close did we come?* 2012, PBS Frontline, n.p., viewed 5 November 2012, <http://www.pbs.org>.

39 M Fackler & C Sang-Hun, *Japan readies in case rocket from North Korea poses threat*, 30 March 2012, The New York Times, New York, viewed 5 November 2012, <http://www.nytimes.com>. Some Japanese media was less assertive, reporting the threats to have been a bluff. (J Hongo, *Vaunted missile shield more for show than protection,* 10 April 2012, The Japan Times Online, Tokyo, viewed 5 November 2012, <http://www.japantimes.co.jp>.)

40 References are:

During May 2013 a total of 17 US officers were stripped of their certifications following a poor assessment of launch operations. In August 2013 a US nuclear missile unit was rated as 'unsatisfactory'. (*US nuclear missile unit deemed unsatisfactory*, 14 August 2013, Space War, Space Media Network, n.p., viewed 24 December 2013, <http://www.spacewar.com>.)

In December 2013 a US general with responsibility for nuclear missiles was reported as having gone on '*a drunken bender*' while in Russia. (*US general went on drunken bender in Russia*, 21 December 2013, Space War, Space Media Network, n.p., viewed 28 December 2013, <http://www.spacewar.com>.)

In January 2014 US missile officers were investigated for possessing illegal drugs. (*Two US missile officers investigated for illegal drugs*, 12 January 2014, Space War, Space Media Network, n.p., viewed 14 January 2014, <http://www.spacewar.com>.)

This was followed by the revelation that 34 officers had been suspended over exam cheating. (*US suspends 34 nuclear missile officers over exam cheating*, 15 January 2014, Space War, Space Media Network, n.p., viewed 16 January 2014 <http://www.spacewar.com>.)

41 *National security space strategy* (Unclassified Summary), 2011, *op. cit.* p. 2, Graph, 'Number of nations and government consortia operating in space'.

42 [Report of the] *Group of governmental experts on transparency and confidence-building measures in outer space activities,* 29 July 2013, reissued 27 September 2013, UN Document A/68/189, Sections IV, V and VI.

43 In 2008 there were 300 000 to 400 000 GPS units employed in agriculture alone. For motor vehicle navigation, approximately 17 000 000 units had been sold in 2008 in the US. (L Grosswald, 'A day without space', *Space Research Today*, vol. 174, April 2009, pp. 3 – 4, recording results of a forum

held under the auspices of the US Chamber of Commerce in October 2008.

The only existing US nation-wide 'interoperable' communications network was critically space dependent; space data and services strongly influenced national security; 90 per cent of military communications systems in Iraq and Afghanistan were transmitted via on non-military satellites. (A *day without space, economic and national security ramifications,* held 16 March 2008. Located at George C Marshall Institute, Arlington, <http://www.marshall.org>, viewed 10 January 2013, <http://www.marshall.org>.)

44 USD 304.31 billion. The space report 2013, Space Foundation, *op. cit.* p. 5. Also, *Global spending on space grew 4% in 2013 to $314 billion*, 21 May 2014, dna India (citing Reuters with no further detail), Diligent Media, Mumbai, viewed 8 July 2014, <http://www.dnaindia.com/world/report-global-spending-on-space-grew-4-in-2013-to-314-billion-1990206>. For fall in government spending, *Global spending on space falls, emerging states are spending more*, 14 February 2014, Reuters, n.p., viewed 8 July 2014. <http://in.reuters.com/article/2014/02/13/space-spending-idINI5NOLI4OB20140213>.

45 Many examples of dual-guidance munitions exist for a range of applications. For example; *Always in control: new guided missiles, programmable munitions enhancing the infantry precision fire effects'*, 2011, Defense Update, Qadima, viewed 11 January 2013, <http://defense-update.com>.

46 *Dead stars could be cosmic 'GPS'*, 9 October 2012, GPS Daily, Space Media Network, n.p., viewed 13 October 2012, <http://www.gpsdaily.com>.

47 *Russia, US to protect satellite navigation at UN level*, 31 October 2013, GPS Daily, n.p., viewed 27 December 2013, <http://www.gpsdaily.com>.

48 RC Solomon & F Flores, *Building trust*, 2003, Oxford Scholarship Online, Oxford, viewed 12 January 2014, <http://www.oxfordscholarship.com>,. YD Luo, 'Building trust in cross-cultural collaborations', *Journal of Management*, vol. 28, iss. 5, 2002, pp. 669 – 694. HC Kelman, 'Building trust among enemies: The central challenge for international conflict resolution', *International Journal of Intercultural Relations*, vol. 29, iss. 6, 2005, pp. 639 – 650.

Chapter 9

1 M Rosenberg, The number of countries in the world, 23 March 2012, The New York Times Company, New York, viewed 3 January 2013, <http://geography.about.com>. Population of member States of the Security Council was accepted as (in millions of people): PRC 1324.7, France 64.4, Britain 61.4, US 304.1 and Russia 142.1. (*Environmental data and statistics*, 2012, The World Bank, Washington DC, viewed 3 January 2013, <http://web.worldbank.org>.) Total global human population as at January 2012 accepted as 6.985 billion. (*World population review*, 2012, US Census Bureau, Washington DC, viewed 3 January 2012, <http://worldpopulationreview.com>.)

Appendix 1

1 Two examples of exoatmosphere/exoatmospheric from US sources are:

'The extended period of flight that the North Korean TBMs [Theatre Ballistic Missiles] *spend in the exo-*

atmosphere provides upper tier TBMD [Theatre Ballistic Missile Defence] *systems ample engagement opportunities....'* Department of Defense, *'Report to Congress on theatre missile defense architecture options for the Asia-Pacific region'*, Accession Number ADA 364008, Washington DC, 1999.

'Raytheon's Exoatmospheric Kill Vehicle (EKV) stands ready to defend the United States against intercontinental ballistic vehicles as a mission-critical component of the Ground-based Midcourse Defense System.' Exoatmospheric kill vehicle, 2013, Raytheon, Tucson, viewed 9 June 2014, <http://www.raytheon.com/capabilities/products/ekv/>.

2 Examples are:

'The test was also aimed at assessing the performance of the interceptor missile's rocket motor system and exoatmospheric kill vehicle, which was designed to collide directly with a target warhead in space to "hit to kill" the target warhead using only the force of the collision.' (*US military conducts successful anti-missile test,* 2 September 2006, China.org.cn, Beijing, viewed 29 December 2010, <http://www.china.org.cn>.)

'In the mid 1980s the Americans started researching the Lightweight Exo-Atmospheric Projectile.' A Kislyakov, *Standard downing of spy satellite,* 2010, RIA Novosti, Moscow, viewed 29 December 2010, <http://en.ria.ru>.

'..an anti-ballistic missile developed to intercept incoming ballistic missiles outside atmosphere (Exoatmosphere)' Prithvi air defence, 13 March 2010, Indian Military, n.p., viewed 29 December 2010.

3 Three examples follow:

'The target [missile] *was then acquired and destroyed in 30 seconds at a height of 48 kilometres. The missile interception was in the exo-atmospheric zone (upper layer of the atmosphere).'* S Ramachandran, *Indian Missile Defence,* 9 December 2006, Asia Times Online, Hong Kong, viewed 29 December 2010,<http://www.atimes.com>.

The US National Aeronautics and Space Administration website hosted a paper by PH Urschel & TH Cox, *Launch condition deviations of reusable launch vehicle simulators in exo-atmospheric zoom climbs,* September 2003. Embedded in the paper was a graph indicating that exo-atmospheric manoeuvres began at an altitude of 200 000 feet (approximately 61 km.). (Located at National Aeronautics and Space Administration, <http://www.nasa.gov >.)

The European concept of exoatmospheric might include higher altitudes. [Astrium] *'Company officials also called on Europe to finance its own ballistic missile defense system, starting with a flight demonstration of an exoatmospheric kill vehicle that would knock out a dummy missile in space at about 300 kilometres in altitude.'* (PB de Selding, *Astrium calls on Europe to fund Soyuz replacement,* 2010, Space News, Imaginova Corporation, Stamford, viewed 27 December 2010, <http://www.spacenews.com>.)

4 For example, a Russian *Progress* resupply vehicle (an autonomous version), visited the International Space Station and, following departure, executed two orbital changes before deploying a *Chibis-M* satellite. (P Harding, *Progress launch,* 30 October 2011, NASA Spaceflight, n.p., viewed 13 March 2013, <http://www.nasaspaceflight.com>.)

5 The PRC has nevertheless claimed a successful test flight of its own space plane, a craft similar to the US X-37B, the test being conducted 'shortly after' the X-37 B test flight of 2010. (*China's space plane succeeds in test fly*, 29 April 2010, Hao Hao Report, n.p., viewed 22 December 2010, <http://www.haohaoreport.com>.)

6 *Secret American space planes to dominate Earth*, 5 November 2002, trans. D Sudakov, Pravda Ru, Moscow, viewed 22 December 2010, < http:///www.english.pravda.ru >.

7 United Press International reported on 14 May 2010 that Iran was concerned at the implications of US space plane activity. (*US space planes worry Iran*, 2010, United Press International, Washington DC, viewed 22 December 2010, <http://www.upi.com>.) Iran's concern was also reported by, Ya Libnan, *US space planes,* 16 May 2010, Lebanon, viewed 22 December 2010, <http://www.yalibnan.com>.

8 For a summary of early US and USSR missile defence developments; Office of Technology Assessment, US Congress, *Strategic Defenses*, Princeton University Press, Princeton, 1986, pp. 45 – 66.

9 AF Woolf, MB Nitikin & PK Kerr, *Arms control and non-proliferation*, Congressional Research Service (document 7.5700, RL 33865), Washington DC, 2010. SA Hildreth & AF Woolf, *Ballistic missile defense and offensive arms reductions*, Congressional Research Service, (document 7.5700, R41251), Washington DC, 2010.

10 UN Charter, Article 51; '*Nothing in the present Charter shall impair the inherent right of individual or collective self-defence if an armed attack occurs against a Member of the UN, until the Security Council has taken measures necessary to maintain international peace and security.*'

11 For a discourse on this capability, illustrated with reference to a superseded satellite constellation, JT Richelson, *America's space sentinels*, University Press of Kansas, Lawrence, 1999.

12 *Japan threatened to deploy missile interceptors,* 28 March 2009, Wall Street Journal Digital Network, New York, viewed 9 December 2010, <http://online.wsj.com>.

13 *North Korea conducts its third nuclear test,* 5 April 2010, The Associated Press, New York, viewed 9 December 2010, <http://www.msnbc.msn.com>.

14 *Statement by Mr Kang Myong hol, Representative of the Democratic People's Republic of Korea at the thematic debate of the first Committee On Cluster 3: Outer Space*, New York, 22 October 2012. Located at; United Nations, Geneva, <http://www.un.org/disarmament>.

15 Security Council, S/RES/2087 (2013) adopted 22 January 2013, *Non-proliferation/Democratic People's Republic of Korea.*

16 '*No force on Earth can block the Korean people all out to defend the independent rights including the right to space development, says Rodong Sinmun...*' (R Sinmun, *DPRK's right to space development for peaceful purposes inviolable,* 28 January 2013, Korean Central News agency of DPRK, Pyongyang, viewed 29 January 2013, <http://www.kcna.co.jp>.)

17 H Bhayana, *International law in the regime of outer space*, R Cambray, Calcutta, 2001.

18 S Sanz Femández de Córdoba, *100 km boundary for astronautics*, n.d., Fédération Aéronautique Internationale, Lausanne, viewed 26 March 2014, <http://www.fai.org>.

19 *Space Activities Act*, 1998, Act 123 of 1998. (Australia)

20 *United States Code of Federal Regulations, Title 32, National Defense*, Chapter V, Department of the Army, Part 578.80, Army aviator badges, the astronaut device.

21 Image courtesy of Windows to the Universe, 2011, National Earth Science Teachers Association, United States, viewed 23 September 2011, <http://www.windows2universe.org>. (Limitations mentioned regarding the diagram were not intended as criticisms. The diagram performed its intended function perfectly and was the clearest illustration found to illustrate the fundamental points made regarding satellites in orbit through the residual atmosphere. Author)

22 United Nations, *Register of space objects*, 2011, United Nations Office of Outer Space Affairs, Vienna, viewed 20 September 2011, <http://www.oosa.unvienna.org>.

23 *Tectonic plate motion*, 2011, National Aeronautics and Space Administration, Washington DC, viewed 17 November 2011, <http://www.nasa.gov.>.

24 *Tides and currents, tidal datums, national tidal datum epoch*, rev. 15 October 2013, National Oceanic and Atmospheric Administration, Washington DC, viewed 18 April 2015, <http://www.tidesandcurrents.noaa.gov>..

25 Image courtesy of National Geospatial Intelligence Agency, *Earth gravitational model (EGM) 2008*, viewed 6 November 2011, <http://earth.info.nga.mil>.

26 *NASA Armstrong Fact Sheet: helios prototype*, 28 February 2014, National Aeronautics and Space Administration, Washington DC, viewed 9 June 2014, <http://www.nasa.gov/centers/armstrong/news/FactSheetsFS-068-DFRC.html >. (reported as 96,863 feet, achieved in August 2001)

27 *Japan sets new balloon altitude record: 53.7 kms – 9/20/13*, September 2013, Stratocat, n.p., viewed 1 February 2014, <http://stratocat.com.ar/news0913e.htm>. (Pre-test publicity had announced the objective was to reach 55 km altitude. - Author)

28 CP Vick, *KH-7 Gambit*, 2011, Global Security, Alexandria, viewed 21 September 2011, <http://www.globalsecurity.org>.

29 National Reconnaissance Office, *The GAMBIT story*, June 1991, (Approved for Release 17 September 2011); National Reconnaissance Office, Chantilly, viewed 30 September 2011, <http://www.nro.mil>.

30 UN General Assembly Resolution 2222 (XXI) of 19 December 1966, [Re] *Treaty on Principles Governing the Activities of States in the Exploration and Use of Outer Space, Including the Moon and Other Celestial Bodies.*

31 UN document A/AC. 105/37 of 14 July 1967. *Study of questions relative to the definition of outer space.*

32 UN document A/AC. 105/935 (2009). *Report of the Legal Subcommittee on its forty-eighth session, held in Vienna from 23 March to 3 April 2009,* p.10, paragraph 55 and, Annex II.

33 List of contracting States, *Member States*, 31 October 2013, International Civil Aviation Organisation, Montreal, viewed 9 June 2014. <http://www.icao.int>.

34 The provisions of the Chicago Convention and those governing the ICAO are less stringent and more flexible than the UN Charter, particularly when the UN is additionally encumbered by the processes such as the consensus agreement arrangement employed by COPUOS. (The representation and voting arrangements for ICAO are at Part II of the *Chicago Convention.*) Conversely, aviation regulation is influenced by the ICAO Council. Its powerful '*States of chief importance in air transport*' include all members of the UN Security Council, implying potential to resist changes not favoured by the most powerful of States.

35 The potential objections here considered were derived from an examination of the consistent opposition by the US to adoption of a definition and delimitation to outer space. An extract from records of the Legal Subcommittee to COPUOS, meeting held 25 March 2010, document COPUOS/LEGAL/T.810, US representative Mr S McDonald, is:

'*As we have stated on previous occasions, the United States is of the view that there is no need to seek a legal definition or delimitation for outer space. The current framework has presented no practical difficulties and indeed activities in outer space are flourishing. Given this situation, an attempt to define of delimit outer space would be an unnecessary and theoretical exercise that could potentially complicate existing activities and might not be able to anticipate continuing technological developments.*'

Appendix 2

1 References are:

J Moltz, *The politics of space security, op. cit.* p. 119, citing PB Stares, *Militarization of space, op. cit.* and W Burrows, *This new ocean,* Random House, New York, 1998.

CKS Chun, *Defending Space, US anti-satellite warfare and space weaponry,* Osprey Publishing, Oxford, 2006, p 33.

PB Stares, *The militarization of space*, Cornell University Press, Ithaca, 1985, p. 108. (Stares cited S Glasstone & P Dolan, *The effects of nuclear weapons*, US Department of Energy and US Department of Defense, Washington DC, 1977, p. 47; and US Congress, Senate, Preparedness Investigating Subcommittee of the Committee on Armed Services, Hearings 88th Congress, 1st Session (1963) Part 1, p. 172.)

The test was confirmed in testimony by G Ullrich, *Statement to 1997 (US) Congressional Hearings, Special Weapons, Nuclear, Chemical, Biological and Missile.* Located at Federation of American Scientists, 2010, Washington DC, <http://www.fas.org>.

2 H Hoerlin, *United States high-altitude test experiences*, 1976, Los Alamos Scientific Laboratory. Located at Federation of American Scientists, <http://www.fas.org>.

3 The author acknowledges material and advice by Professor John Kennewell, Curtin University and Australian Space Academy, as providing the theme for this appendix. The term *Nuclear-Detonation-Induced Radiation Effects* (NDRE) possibly originated with the author although it conceivably had been employed by others. One valuable reference work consulted was, S Glasstone & PJ Dolan, *The effects of nuclear weapons,* 3rd edn, United States Department of Defense and United States Department of Energy, Washington DC, 1977.

4 *Developing threats: Electro-magnetic pulses (EMP), Non-Nuclear EMP*, Defence Committee, Parliament of the United Kingdom, London, viewed 12 October 2013, <http://www.publications.parliament.uk>.

5 Figure courtesy of Professor John Kennewell, Curtin University and Australian Space Academy, 2013.

6 G Ullrich, *Statement to 1997 (US) Congressional Hearings, Special Weapons, Nuclear, Chemical, Biological and Missile, op. cit.* E Savage, J Gilbert & W Radasky, *The early time (E1) high-altitude electromagnetic pulse (HEMP) and its impact on the US power grid*, Metatech Corporation, Goleta, 2010. *Developing threats: Electro-magnetic pulses (EMP), Non-Nuclear EMP, op. cit.*

7 RN Ghose, *Environment and system hardness design*, Don White Consultants, Gainesville, 1983, p. 2.7. J Davis & RS Murch (sponsors), *High altitude nuclear detonations (HAND) against low earth orbit satellites ("HALOES")*, April 2001, Defense Threat Reduction Agency, Advanced Systems and Concepts Office. Located at Federation of American Scientists, <http://www.fas.org>. G Ullrich, testimony, *op. cit.*

8 LW Ricketts, *Fundamentals of nuclear hardening of electronic equipment*, Wiley-Interscience, New York, 1972, p. 115, *et ceq.*

9 Advice from Professor JA Kennewell, Curtin University and Australian Space Academy, March 2014.

10 D Wright, L Grego and L Gronlund, *The Physics of Space Security, A reference manual,* American Academy of Arts and Sciences, Cambridge, 2005, pp. 138 - 139.

11 J Davis & RS Murch (sponsors), *High altitude nuclear detonations (HAND), op.cit.*

12 For a list, possibly incomplete, of US satellites hardened to resist 'prompt, high-dose effects of nuclear explosions,' J Davis & RS Murch (sponsors), *High altitude nuclear detonations (HAND) against low earth orbit satellites (HALOES), op. cit.* p. 29.

13 R Locoe, 'Designing radiation hardened CMOS microelectronic components at commercial foundries; space and terrestrial radiation environments and device and circuit techniques to mitigate radiation effects,' *Integrated Reliability Workshop Final Report,* 2005, IEEE, international, <http://www.ieeexplore.ieee.org>. R Tu, G Lum, P Pavan, *et al. Simulating total- dose radiation effects on circuit behaviour*, 1994, Reliability Physics Symposium, 32nd annual proceedings, pp. 344 – 350. Located at, <http://www.ieeexplore.ieee.org>. A Keys & M Watson, *Radiation hardened*

electronics for extreme environments, c. 1997, National Aeronautics and Space Administration, Washington DC, <http://www.nasa.gov/archive>.

14 J Davis & RS Murch (sponsors), *High altitude nuclear detonations, op. cit.* p. 4.

15 JS Foster, E Gjelde, WR Graham, *et al. Report of the Commission to Assess the Threat to the United States from Electromagnetic Pulse (EMP) Attack, Critical National Infrastructures*, April 2008, pp. 58 – 171. Located at, EMP Commission, McLean, <http://www.empcommission.org >.

16 For a summarised discussion; D Wright, L Grego & L Gronlund, *op. cit.* pp. 138 -139.

www.ingramcontent.com/pod-product-compliance
Ingram Content Group UK Ltd.
Pitfield, Milton Keynes, MK11 3LW, UK
UKHW061437070726
13610UKWH00019B/108